THE HISTORY AND FUTURE OF MEGALODONS

ATHARVA S. MOHITE

Contents

Preface

This book throws light on the Theory regarding the evolution of megalodons. The book strives to elaborate upon the two species of shark that are considered. The book also comprises another theory regarding the evolution of megalodons into modern day great whites. The book also talks about a hypothetical design of an incubator that can be used for the incubation and growth of embryos.

CHAPTER ONE

Introduction to Megalodons........

According to Britannica, Megalodon,(Carcharocles megalodon) was a member of a defunct species of shark(Otodontidae) which are considered to be the largest species of shark, as well as the largest fish, that have ever lived. Fossils related to megalodon have been set up dating from the early Miocene Period(which began 23.03 million times agone) to the end of the Pliocene Period (2.58 million times agone). The word megalodon, a emulsion of Greek root words, means " giant tooth. " Megalodon (Carcharocles megalodon) had a 3- cadence(9.8- bottom) bite periphery, which is several times larger than that of ultramodern great white shark(Carcharodon carcharias). Reactionary remains of megalodon have been set up in shallow tropical and temperate swell along the plages and international shelf regions of all mainlands except Antarctica. During the early and middle corridor of the Miocene Period(which lasted from 23 million to5.3 million times agone), large seaways separated North America from South America and Europe and Asia from Africa and the Middle East, which probably eased movement from one sea receptacle to another. Throughout the Miocene, megalodon distribution expanded from pockets located in the Caribbean and Mediterranean swell, in the Bay of Bengal, and along the beachfronts of California and southern Australia to encompass waters off the beachfronts of northern Europe, South America, southern Africa, New Zealand, and east Asia. During the Pliocene Period, still, the megalodon's geographic range contracted significantly, and it was defunct by the end of the time. Megalodon was the largest fish ever known, a designation grounded on discoveries of hundreds of reactionary teeth and a sprinkle of chines. Tooth- shape parallels between megalodon and ultramodern great white shark(Carcharodon carcharias) suggest that the two species may have been close cousins, and therefore megalodon likely recalled that species in appearance — that is, as a big torpedo- shaped fish with a conical conk, large pectoral and rearward fins, and a strong crescent- shaped tail. Estimates of body length are calculated using the statistical relationship between the size of megalodon's reactionary teeth and the teeth and body mass of ultramodern white shark and other living cousins. This data suggests that mature adult megalodons had a mean length of 10.2 measures (about33.5 bases), the largest samples measuring17.9 measures (58.7 bases) long. Some scientists, still, contend that the largest forms may have measured up to 25 measures(82 bases) long. Studies estimate that adult body mass ranged from roughly 30 metric tons(1 metric ton = 1,000 kg; about 66,000 pounds) to further than 65 metric tons(about 143,000 pounds), adult females being larger(in both length and mass) than adult males. Megalodon teeth are analogous to those of ultramodern white shark in that they're triangular, saw- toothed, and symmetrical. They differ from ultramodern white shark teeth in that they're larger and thicker, the serrations on each tooth do in regular intervals, and they retain a bourlette(a darker, badge- shaped region near the tooth's root). The largest extant megalodon tooth measures17.8 cm(6.9 elevation) in length, nearly three times longer than those of ultramodern white shark(which are generally about5.4 cm(2.1 elevation) long). In addition, the megalodon held a ferocious bite; its bite periphery was 3 measures(about9.8 bases), several times larger than the bite periphery of average- sized white shark.

Megalodon is allowed to have managed its body temperature like that of ultramodern white shark, in that it wasn't simply cold- thoroughbred like utmost fish. White shark induce heat through the compression of their swimming muscles, and this heat raises the temperature of corridor of the shark's body above that of the girding water, an adaption called indigenous endothermy(which is a type of warm- bloodedness). This adaption might have allowed megalodon to syncope and quest in colder waters, giving it exclusive access to prey in those locales. Reproduction and territoriality Megalodon is allowed to have produced live youthful. It isn't known, still, whether the species was ovoviviparous(in which eggs are retained within the mama until they door) or viviparous(in which fertilized embryos decide nonstop aliment from the mama). Estimates of body size using juvenile teeth suggest that recently produced youthful may have been at least 2 measures(6.6 bases) in length. Many details are known about megalodon courting, but the species appears to have used nurseries for its youthful. A 2010 study linked a megalodon nursery along the Panamanian seacoast, which was characterized by the presence of juvenile teeth from colorful stages of life. Scientists posit that this shallow warm- water nursery handed youthful megalodons with access to a different array of lower, more abundant prey and enabled grown-ups to more block attacks from other raptorial shark species, similar as hammerhead shark. As the youthful shark grew aged, it was allowed that they would make raids into deeper water to pursue larger creatures. Little is known about how individualities dispersed after they progressed. Since megalodon is allowed to have enthralled an ecological niche analogous to that of the white shark, some studies have assumed that megalodon likely ranged over areas similar in size to the range of ultramodern white shark — about 1,000 square km(386 square long hauls). Predators and prey In addition to being the world's largest fish, the megalodon may have been the largest marine bloodsucker that has ever lived.(Basilosaurids and pliosaurs may have been just as large.) Megalodon was an apex bloodsucker, or top carnivore, in the marine surroundings it inhabited(see also cornerstone species). It feed upon fish, baleen jumbos, toothed jumbos(similar as ancestral forms of ultramodern sperm jumbos, dolphins, and killer jumbos), sirenians(similar as dugongs and manatees), and seals. The youthful likely sought out lower prey, while grown-ups hunted larger jumbos. Mature megalodons probably didn't have any predators, but recently produced and juvenile individualities may have been vulnerable to other large raptorial shark, similar as great hammerhead shark(Sphyrna mokarran), whose ranges and nurseries are allowed to have lapped with those of megalodon from the end of the Miocene and throughout the Pliocene. Paleontology important debate continues to compass the taxonomy and elaboration of megalodon(Carcharocles megalodon(in some groups Otodus megalodon)), as well as its relationship to ultramodern white shark(Carcharodon carcharias). Megalodon was first described in 1835 by

Swiss- born American naturalist, geologist, and schoolteacher Louis Agassiz, who named the species Carcharodon megalodon. Megalodon would be known by this scientific name until the late 1990s when a growing group of scientists placed it in the rubric Carcharocles. Although some paleontologists assert that megalodon and ultramodern white shark evolved within the same lineage(Carcharodon in the shark family Lamnidae) grounded on their saw-toothed teeth, others classify megalodon within the lineage of megatooth shark(Otodontidae) whose origins trace back to the Cretaceous Period(145 million to 66 million times agone). A tooth-analysis study performed in 2012 suggested that the ultramodern white shark evolved from the lamnid shark(Lamnidae) some 5 million times ago during the late Miocene and early Pliocene ages. The study notes that the pattern of serrations and other parallels in tooth structure between megalodon and ultramodern white sharks could be a product of coincident elaboration(wherein analogous characters evolve singly in different lineages). Extermination Megalodon's geographic distribution expanded throughout the Miocene but contracted during the Pliocene as populations declined. Originally, scientists believed that the decline was due to swings in sea temperatures related to climate change, conceivably caused by the ending of the seaway separating North America and South America about 3 million times agone, which veered sea currents and caused other changes in sea rotation.

By 2016, studies had shown that megalodon's geographic distribution didn't increase appreciably during warm ages or drop appreciably during cold ages, suggesting that the species' demise wasn't dependent on climatic changes alone. These studies suggested that shifting food-chain dynamics may have been the primary factor in megalodon's demise, as the vacuity of its primary food source, baleen jumbos, dropped, and the figures of its challengers — lower raptorial shark(similar to the great white shark, Carcharodon carcharias) and jumbos(similar as members of the killer Goliath genus Orcinus) increased. Several studies note that authentic reactionary teeth attributed to megalodon don't do later than the Pliocene-Pleistocene boundary(2.58 million times agone), suggesting that megalodon failed out around that time. still, one study, which has noted data problems associated with the others, has questioned the validity of reactionary teeth dating to the late Pliocene. It reports rather that validated reactionary data suggests that megalodon failed out near the end of the first stage of the Pliocene, the Zanclean Stage(about 3.6 million times agone).

CHAPTER TWO

Introduction to Modern day Great Whites

As per an article published by Wikipedia. The great white shark(Carcharodon carcharias), also known as the white shark, white pointer, or simply great white, is a species which can be set up in the littoral face waters of all the major abysses. It's notable for its size, with larger lady individualities growing to6.1 m(20 ft) in length and 1,905 – 2,268 kg(4,200 – 5,000 lb) in weight at maturity.(3)(4)(5) still, utmost are lower; males measure3.4 to4.0 m(11 to 13 ft), and females measure4.6 to4.9 m(15 to 16 ft) on normal.(4)(6) According to a 2014 study, the lifetime of great white shark is estimated to be as long as 70 times or further, well above former estimates,(7) making it one of the longest lived cartilaginous fishes presently known.(8) According to the same study, manly great white shark take 26 times to reach sexual maturity, while the females take 33 times to be ready to produce seed. (9) Great white shark can swim at pets of 25 km/ hr(16 mph)(10) for short bursts and to depths of 1,200 m(3,900 ft). (11) The great white shark is an apex bloodsucker, as it has no given natural predators other than, on veritably rare occasions, the orca. (12) It's arguably the world's largest-known extant macro predatory fish, and is one of the primary predators of marine mammals, up to the size of large baleen jumbos. This shark is also known to prey upon a variety of other marine creatures, including fish, and seabirds. It's the only known surviving species of its rubric Carcharodon, and is responsible formorer recorded mortal bite incidents than any other shark. (13)(14) The species faces multitudinous ecological challenges which has redounded in transnational protection. The International Union for Conservation of Nature lists the great white shark as a vulnerable species,(1) and it's included in excursus II of CITES. (15) It's also defended by several public governments, similar to Australia(as of 2018). (16) Due to their need to travel long distances for seasonal migration and extremely demanding diet, it isn't logistically doable to keep great white sharks in prison; because of this, while attempts have been made to do so in the history, there are no given fences in the world believed to house a live instance. (17)

The novel Jaws by Peter Benchley and its posterior film adaption by Steven Spielberg depicted the great white shark as a ferocious man-eater. Humans aren't the preferred prey of the great white shark,(18) but the great white is nonetheless responsible for the largest number of reported and linked fatal unprovoked shark attacks on humans, although this happens veritably infrequently(generally smaller than 10 times a time encyclopedically).(19)(20)

Etymology and naming history

The name 'great white shark' likely comes from the shark's size, as well as the white underside exposed on beached sharks.

The English name 'white shark' and its Australian variant 'white pointer'[23] are thought to have come from the shark's stark white underside, a characteristic feature most noticeable in beached sharks lying upside down with their bellies exposed.[24] Colloquial use favors the name 'great white shark', perhaps because 'great' stress the size and prowess of the species.[25] Another reason might be that "white shark" was a term historically used to describe The oceanic white-tipped shark, and thus, being much larger than the latter shark, it was named "Great" as the "white" part of its name was already used for another shark, which was subsequently referred to as the "lesser white shark". Most scientists prefer 'white shark', due to the fact the name "lesser white shark" is no longer used.[25] Some use 'white shark' to refer to all members of the Lamnidae.[22]

The scientific genus names *Carcharodon* means "jagged tooth", a reference to the large serrations that appear in the shark's teeth. Broken down, it is a portmanteau of two Ancient Greek words. The prefix *carchar-* is derived from καρχαρίας (*kárkharos*), which means "jagged" or "sharp". The suffix *-odon* is a Romanization of ὀδών (*odṓn*), which translates to "tooth". The specific name*carcharias* is a Latinization of καρχαρίας (*karkharías*), the Ancient Greek word for shark.[21] TheGreat white was one of the species originally described by Carl Linnaeus in his 1758 *10th edition of Systema Naturae*, in which it was identified as an _amphibian_ and assigned the scientific name _Squalus carcharias_, _Squalus_ being the genus that he placed all sharks in.[26] By the 1810s, it was recognized that the shark should be placed in a new genus but it was not until 1838 that Sir Andrew Smith coined the name _Carcharodon_ as the new genus.[27]

There have been many attempts to describe and classify the great white before Linnaeus. One of its foremost mentions in literature as a distinct type of beast appears in Pierre Belon's 1553 book De aquatic bus brace, cum minibus announcement vivam ipsorum effigiem quoad ejus fieri potuit, announcement amplissimum cardinalem Castilioneum. In it, he illustrated and described the shark under the name Canis carcharias grounded on the jagged nature of its teeth and its alleged parallels with tykes. (a) Another name used for the great white around this time was Lamia, first chased by Guillaume Rondelet in his 1554 book Libri de Piscibus Marinis, who also linked it to the fish that swallowed the prophet Jonah in biblical textbooks.(28) Linnaeus honored both names as former groups. (26) Molecular timepiece studies published between 1988 and 2002 determined the closest living relation of the great white to be the Mako shark of the rubric Isurus, which diverged sometime between 60 to 43 million times agone. (30)(31) Tracing this evolutionary relationship through fossil substantiation, still, remains subject to further paleontological study. (31) The original thesis of the great white shark's origin held that it's an assignee of a lineage of mega-toothed sharks, and is nearly related to the neolithic megalodon. (31)(32) These sharks were vastly large, with megalodon attaining an estimated length of over to14.2 – 16 m(47 – 52 ft). (33)(34) parallels between the teeth of a great white and mega-too shark, similar as large triangular shapes, serrated blades, and the presence of dental bands, led to the primary substantiation of a close evolutionary relationship. As a result, scientists classified the ancient forms under the rubric Carcharodon. Although sins in the thesis were, similar to query over exactly which species evolved into the ultramodern great white and multiple gaps in the reactionary record, paleontologists were suitable to chart the academic lineage back to a 60-million-time-old shark known as Cretalamna as the common ancestor of all shark within the Lamnidae. (30)(32)

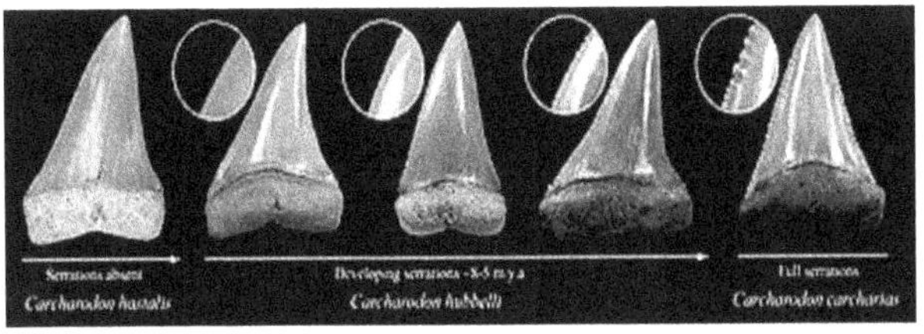

Illustrated elaboration from C. hastalis to C. carcharias still, it's now understood that the great white shark holds near ties to the Mako shark and is descended from a separate lineage as a chronospecies unconnected to the mega-toothed shark. (31) This was proven with the discovery of a transitional species that connected the great white to an unserrated shark known as Carcharodon hastalis. (35)(36) these transitional species, which was named Carcharodon Hubbell in 2012, demonstrated a mosaic of evolutionary transitions between the great white and. hastalis, videlicet the gradational appearance of serrations,(35) in a span of between 8 to 5 million times agone(37) The progression ofC. Hubbell characterized shifting diets and niches; by 6.5 million times agone, the serrations were developed enough for C. hubbelli to handle marine mammals. (35) Although both the great white and C. hastalis were known worldwide,(31)C. Hubbell is primarily set up in California, Peru, Chile, and girding littoral deposits,(38) indicating that the great white had Pacific origins. (35) C. hastalis continued to thrive alongside the great white until its last appearance around one million times agone (39) and is believed to have conceivably got several fresh species, including Carcharodon sub serratus (31)(35) and Carcharodon plicatilis. (31) still, Yun argued that the tooth reactionary remains of C. hastalis and Great white" have been proved from the same deposits, hence the former can not be a conspecific ancestor of the ultimate."(40) He also blamed the C. hastalis" morphotype has noway been tested through phylogenetic analyses," and denoted that as of 2021, the argument that the ultramodern Carcharodon lineage with narrow, saw-toothed teeth evolved from C. hastalis with broad, unserrated teeth is uncertain.(40) Tracing beyond C. hastalis, another prevailing thesis proposes that the great white and Mako lineages participated a common ancestor in a primitive Mako- suchlike species.(41) The identity of this ancestor is still batted, but an implicit species includes Isurolamna inflata, which lived between 65 to 55 million times agone.

It's hypothecated that the great white and Mako lineages resolve with the rise of two separate descendants, the one representing the great white shark lineage being Macrorhizodus precursor.(41)(42)

Distribution and habitat

Shark in Guadalupe Island Biosphere Reserve, Mexico Great white sharks live in nearly all littoral and coastal waters which have water temperatures between 12 and 24 °C(54 and 75 °F), with lesser attention in the United States(Northeast and California), South Africa, Japan, Seaia, Chile, and the Mediterranean including sea of Marmara and Bosphorus. (43)(44) One of the thick-known populations is set up around Dyer Island, South Africa. (45) The great white is an epipelagic fish, observed substantially in the presence of rich game, similar as fur seals(Arctocephalusssp.), sea Napoleons, cetaceans, other shark, and large bony fish species. In the open sea, it has been recorded at depths as great as 1,200 m(3,900 ft). (11) These findings challenge the traditional notion that the great white is a littoral species. (11) According to a recent study, California great whites have migrated to an area between the Baja California Peninsula and Hawaii known as the White Shark Café to spend at least 100 days before migrating back to Baja. On the trip out, they swim sluggishly and dive down to around 900 m(3,000 ft). After they arrive, they change geste and do short dives to about 300 m(980 ft) for over ten twinkles. Another white shark that was tagged off the South African seacoast swam to the southern seacoast of Australia and back within the time. An analogous study tracked a different great white shark from South Africa swimming to Australia's northwestern seacoast and back, a trip of 20,000 km(12,000 mi; 11,000 nmi) in under nine months. (46) These compliances argue against traditional propositions that white shark are littoral territorial predators, and open up the possibility of commerce between shark populations that were preliminarily allowed to have been separate. The reasons for their migration and what they do at their destination are still unknown. Possibilities include seasonal feeding or lovemaking. (47) In the Northwest Atlantic, the white shark populations off the New England seacoast were nearly canceled due to over-fishing .(48) In recent times, the populations have grown greatly,(49) largely due to the increase in seal populations on Cape Cod, Massachusetts since the enactment of the Marine Mammal Protection Act in 1972. (50) Presently veritably little is known about the stalking and movement patterns of great whites off Cape Cod, but ongoing studies hope to offer sapience into this growing shark population. (51) The Massachusetts Division of Marine Fisheries(part of the Department of Fish and Game) began a population study in 2014; since 2019, this exploration has concentrated on how humans can avoid conflict with sharks.(52) A 2018 study indicated that white sharks prefer to congregate deep in anticyclonic orbits in the North Atlantic Sea. The shark studied tended to favor the warm- water orbits, spending the day hours at 450 meters and coming to the face at night. (53)

Anatomy and appearance

Upper Teeth

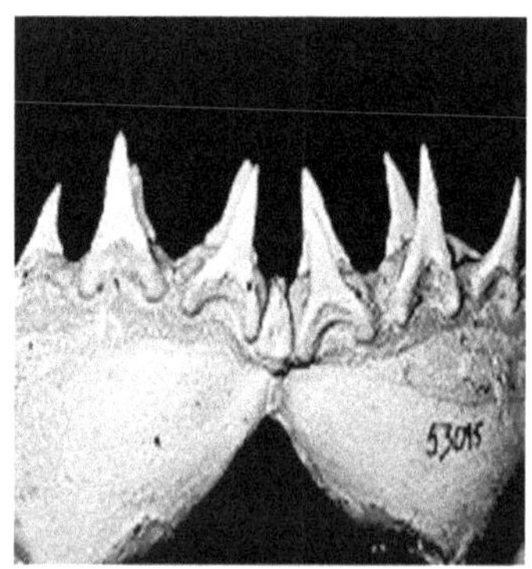

Lower Teeth

Great white Shark shell The great white shark has a robust, large, conical con. The upper and lower lobes on the tail fin are roughly the same size which is analogous to some mackerel shark. A great white displays fight shadowing, by having a white underpart and a slate rearward area(occasionally in a brown or blue shade) that gives an overall mottled appearance. The achromatism makes it delicate for prey to spot the shark because it breaks up the shark's figure when seen from the side. From over, the darker shade blends with the sea, and from below it exposes a minimum figure against the sun. Leucism is extremely rare in this species, but has been proven in one great white shark(a doggy that washed ashore in Australia and failed). (54) Great white sharks, like numerous other sharks, have rows of saw-toothed teeth behind the main bone, ready to replace any that break off. When the shark mouthfuls, it shakes its head side-to-sidesee, helping the teeth saw off large gobbets of meat. (55) Great white shark, like other mackerel shark, have larger eyes than other shark species in proportion to their body size. The iris of the eye is a deep blue rather of black.(56)

Size

Specimen caught off Cuba in 1945 which was allegedly6.4 m(21 ft) long and counted an estimated 3,175 – 3,324 kg(7,000 – 7,328 lb).(57)(58) latterly studies proved this instance to be in the normal size range, at around4.9 m(16 ft) in length.(4) In great white shark, sexual dimorphism is present, and females are generally larger than males. manly great whites on average measure3.4 to4.0 m(11 to 13 ft) long, while females at4.6 to4.9 m(15 to 16 ft).(6) Grown-ups of this species weigh 522 – 771 kg(1,151 – 1,700 lb) on normal;(59) still, mature females can have an average mass of 680 – 1,110 kg(1,500 – 2,450 lb).(4) The largest females have been vindicated up to6.1 m(20 ft) in length and an estimated 1,905 kg(4,200 lb) in weight,(4) maybe up to 2,268 kg(5,000 lb).(5) The maximum size is subject to debate because some reports are rough estimations or enterprises performed under questionable circumstances.(60) Among living cartilaginous fish, only the Goliath shark(Rhincodon typus), the reposing shark(Cetorhinus maximus) and the giant manta ray shaft(Manta birostris), in that order, are on normal larger and heavier. These three species are generally relatively amenable in disposition and given to passively sludge- feeding on veritably small organisms.(59) This makes the great white shark the largest extant macro raptorial fish. Great white shark are at around1.2 m(3.9 ft) when born, and grow about 25 cm(9.8 in) each time.(61) According toJ.E. Randall, the largest white shark reliably measured was a5.94 m(19.5 ft) individual reported from Ledge Point, Western Australia in 1987.(62) Another great white instance of analogous size has been vindicated by the Canadian Shark Research Center A womanish caught by David McKendrick of Alberton, Prince Edward Island, in August 1988 in the Gulf ofSt. Lawrence off Prince Edward Island. This womanish great white was6.1 m(20 ft) long.(4) still, there was a report considered dependable by some experts in the history, of a larger great white shark instance from Cuba in 1945.(58)(63)(64)(65) This instance was reportedly6.4 m(21 ft) long and had a body mass estimated at 3,324 kg(7,328 lb).(58)(64) still, latterly studies also revealed that this particular instance was actually around4.9 m(16 ft) in length, a instance in the average maximum size range.(4) The largest great white honored by the International Game Fish Association(IGFA) is one caught by Alf Dean in south Australian waters in 1959, importing 1,208 kg(2,663 lb).(60) exemplifications of large unconfirmed great whites A number of veritably large unconfirmed great white shark samples have been recorded.(66) For decades, numerous ichthyological workshop, as well as the Guinness Book of

World Records, listed two great white shark as the largest individualities In the 1870s, a10.9 m(36 ft) great white captured in southern Australian waters, near Port Fairy, and an11.3 m(37 ft) shark trapped in a herring block in New Brunswick, Canada, in the 1930s. still, these measures weren't attained in a rigorous, scientifically valid manner, and experimenters have questioned the trustability of these measures for a long time, noting they were much larger than any other directly reported sighting. latterly studies proved these dubieties to be well- innovated. This New Brunswick shark may have been a misapplied basking shark, as the two have analogous body shapes. The question of the Port Fairy shark was settled in the 1970s whenJ.E. Randall examined the shark's jaws and" set up that the Port Fairy shark was of the order of 5 m(16 ft) in length and suggested that a mistake had been made in the original record, in 1870, of the shark's length".(62)

Adaptations

Great white shark, like all other shark, have an redundant sense given by the ampullae of Lorenzini which enables them to descry the electromagnetic field emitted by the movement of living creatures. Great whites are so sensitive they can descry variations of half a billionth of a volt. At close range, this allows the shark to detect indeed immobile creatures by detecting their heartbeat.(97) utmost fish have a less- developed but analogous sense using their body's side line.(98)

Ecology and behavior

Shark smelling into the fish head teaser bait next to a cage in False Bay, South Africa To more successfully quest fast and nimble prey similar as sea lions, the great white has acclimated to maintain a body temperature warmer than the girding water. One of these acclimations is a" rete mirabile"(Latin for" awful net"). This close web- suchlike structure of modes and arteries, located along each side side of the shark, conserves heat by warming the cooler arterial blood with the venous blood that has been warmed by the working muscles. This keeps certain region of the body(particularly the stomach) at temperatures up to 14 °C(25 °F)(99) above that of the girding water, while the heart and gills remain at sea temperature. When conserving energy, the core body temperature can drop to match the surroundings. A great white shark's success in raising its core temperature is an illustration of gigantothermy. thus, the great white shark can be considered an endothermic poikilotherm or mesotherm because its body temperature isn't constant but is internally regulated.(55)(100) Great whites also calculate on the fat and oils stored within their livers for long- distance migrations across nutrient-poor areas of the abysses.(101) Studies by Stanford University and the Monterey Bay Aquarium published on 17 July 2013 revealed that in addition to controlling the shark' buoyancy, the liver of great whites is essential in migration patterns. Sharks that sink briskly during drift dives were revealed to use up their internal stores of energy hastily than those which sink in a dive at further tardy rates.(102) toxin from heavy essence seems to have little negative effects onGreat whites. Blood samples taken from forty- three individualities of varying size, age and sex off the South African seacoast led by biologists from the University of Miami in 2012 indicates that despite high levels of mercury, lead, and arsenic, there was no sign of raised white blood cell count and granulate to lymphocyte rates, indicating the shark had healthy vulnerable systems. This discovery suggests a preliminarily unknown physiological defence against heavy essence poisoning. Great whites are known to have a propensity for" tone- mending and avoiding age- related affections".(103) suck force A 2007 study from the University of New South Wales in Sydney, Australia, used CT reviews of a shark's cranium and computer models to measure the shark's maximum bite force. The study reveals the forces and behaviours its cranium is acclimated to handle and resolves contending propositions about its feeding behavior.(104) In 2008, a platoon of scientists led by Stephen Wroe conducted an trial to determine the great white shark's jaw power and findings indicated that a instance gathering 3,324 kg(7,328 lb) could ply a bite force of 18,216 newtons(4,095 lbf).(64)

A shark turns onto its reverse while hunting tuna bait This shark's geste and social structure is complex.(105) In South Africa, white shark have a dominance scale depending on the size, coitus and squatter's rights Females dominate males, larger shark dominate lower shark, and residers dominate beginners. When hunting, great whites tend to separate and resolve conflicts with rituals and displays. White shark infrequently resort to combat although some individualities have been set up with bite marks that match those of other white shark. This suggests that when a great white approaches too nearly to another, they reply with a warning bite. Another possibility is that white shark suck to show their dominance. Data acquired from beast- borne telemetry receivers and published in 2022 via the journal Royal Society Publishing suggests that individual great whites may associate so that they can inadvertently partake information on the whereabouts of prey or the position of the remains of creatures that can be scavenged. As biologging can help to reveal social habits, it allows a better understanding to be made in unborn studies regarding the full extent of social relations in large marine creatures, including the great white shark.(106) The great white shark is one of only a many shark known to regularly lift its head above the sea face to aspect at other objects similar as prey. This is known as asset- hopping. This geste has also been seen in at least one group of blacktip reef shark, but this might be learned from commerce with humans(it is theorized that the shark may also be suitable to smell more this way because smell peregrination through air briskly than through water). White shark are generally veritably curious creatures, display intelligence and may also turn to fraternizing if the situation demands it. At Seal Island, white shark have been observed arriving and departing in stable" clans" of two to six individualities on a monthly base. Whether clan members are related is unknown, but they get along peacefully enough. In fact, the social structure of a clan is presumably most aptly compared to that of a shark pack; in that each member has a easily established rank and each clan has an nascence leader. When members of different clans meet, they establish social rank nonviolently through any of a variety of relations.(107)

Diet

Great white shark are rapacious and prey upon fish(e.g. tuna, rays, other shark),(107) cetaceans(i.e., dolphins, porpoises, whales), pinnipeds(e.g. seals, fur seals,(107) and sea Napoleons), sea turtles,(107) sea otters(Enhydra lutris) and seabirds.(108) Great whites have also been known to eat objects that they're unfit to digest. Juvenile white shark generally prey on fish, including other elasmobranchs, as their jaws aren't strong enough to repel the forces needed to attack larger prey similar as pinnipeds and cetaceans until they reach a length of 3 m(9.8 ft) or further, at which point their jaw cartilage mineralizes enough to repel the impact of smelling into larger prey species.(109) Upon approaching a length of nearly 4 m(13 ft), great white shark begin to target generally marine mammals for food, though individual shark feel to specialize in different types of prey depending on their preferences.(110)(111) They feel to be largely opportunistic.(112)(113) These shark prefer prey with a high content of energy-rich fat. shark expert Peter Klimley used a rod- and- reel rig and combed carcasses of a seal, a pig, and a lamb from his boat in the South Farallons. The shark attacked all three baits but rejected the lamb corpse.(114) Off of Seal Island, False Bay in South Africa, the shark ambush brown fur seals(Arctocephalus pusillus) from below at high speeds, hitting the sealmid-body. They achieve high speeds that allow them to fully transgress the surface of the water. The peak burst speed is estimated to be over 40 km/ h(25 mph).(115) They've also been observed chasing prey after a missed attack. Prey is generally attacked at the face.(116) shark attacks do most frequently in the morning, within two hours of daylight, when visibility is poor. Their success rate is 55 in the first two hours, falling to 40 in late morning after which stalking stops.(107) Off California, shark incapacitate northern giant seals(Mirounga angustirostris) with a large bite to the hindquarters(which is the main source of the seal's mobility) and stay for the seal to bleed to death. This fashion is especially used on adult manly giant seals, which are generally larger than the shark, ranging between 1,500 and 2,000 kg(3,300 and 4,400 lb), and are potentially dangerous adversaries.(117)(118) utmost generally however, juvenile giant seals are the most constantly eaten at giant seal colonies.(119) Prey is typically attackedsub-surface. Harbor seals(Phoca vitulina) are taken from the face and dragged down until they stop floundering. They're also eaten near the bottom. California sea Napoleons(Zalophus californianus) are ambuscaded from below and struckmid-body before being dragged and eaten.(120)

Great white Shark near Gansbaai, showing upper and lower teeth In the Northwest Atlantic develop great whites are known to feed on both harbor and slate seals.(50)

Unlike grown-ups, juvenile white shark in the area feed on lower fish species until they're large enough to prey on marine mammals similar as seals.(121) White shark also attack dolphins and porpoises from over, behind or below to avoid being detected by their echolocation. Targeted species include dim dolphins(Sagmatias obscurus),(67) Risso's dolphins(Grampus griseus),(67) bottlenose dolphins(Tursiopsssp.),(67)(122) humpback dolphins(Sousassp.),(122) harbour porpoises(Phocoena phocoena),(67) and Dall's porpoises(Phocoenoides dalli).(67) Groups of dolphins have sometimes been observed defending themselves from shark with mobbing geste .(122) White shark predation on other species of small cetacean has also been observed. In August 1989, a1.8 m(5.9 ft) juvenile manly pygmy sperm whale(Kogia breviceps) was set up stranded in central California with a bite mark on its caudal peduncle from a great white shark.(123) In addition, white shark attack and prey upon beaked whales.(67)(122) Cases where an adult Stejneger's beaked whale(Mesoplodon stejnegeri), with a mean mass of around 1,100 kg(2,400 lb),(124) and a juvenile Cuvier's beaked whale(Ziphius cavirostris), an individual estimated at 3 m(9.8 ft), were hunted and killed by great white shark have also been observed.(125) When hunting sea turtles, they appear to simply suck through the carapace around a flipper, prostrating the turtle. The heaviest species of bony fish, the oceanic sunfish(Mola mola), has been set up in great white shark stomachs.(112) Whale cadavers comprise an important part of the diet of white shark. still, this has infrequently been observed due to jumbos dying in remote areas. It has been estimated that 30 kg(66 lb) of Goliath blubber could feed a4.5 m(15 ft) white shark for1.5 months. Detailed compliances were made of four Goliath cadavers in False Bay between 2000 and 2010. shark were drawn to the corpse by chemical and odor discovery, spread by strong winds. After originally feeding on the Goliath caudal peduncle and strike, the shark would probe the corpse by sluggishly swimming around it and mouthing several corridor before opting a blubber-rich area. During feeding bouts of 15 – 20 seconds the shark removed meat with side headshakes, without the defensive optical gyration they employ when attacking live prey. The shark were constantly observed regurgitating gobbets of blubber and incontinently returning to feed, conceivably in order to replace low energy yield pieces with high energy yield pieces, using their teeth as mechanoreceptors to distinguish them. After feeding for several hours, the shark appeared to come sleepy, no longer swimming to the face; they were observed mouthing the corpse but supposedly unfit to suck hard enough to remove meat, they would rather bounce off and sluggishly sink. Up to eight shark were observed feeding contemporaneously, hitting into each other without showing any signs of aggression; on one occasion a shark accidentally smelled the head of a neighboring shark, leaving two teeth bedded, but both continued to feed untroubled. lower individualities floated around the corpse eating gobbets that drifted down. Surprisingly for the area, large figures of shark over five metres long were observed, suggesting that the largest shark change their geste to search for jumbos as they lose the manoeuvrability needed to hunt seals. The probing platoon concluded that the significance of Goliath cadavers, particularly for the largest white shark, has been undervalued.(126) A shark scavenging on a Goliath corpse in False Bay, South Africa In another proved incident, white shark were observed scavenging on a Goliath corpse alongside barracuda shark.(

127) In 2020, marine biologists Sasha Dines and Enrico Gennari published a proved incident in the journal Marine and Freshwater Research of a group of great white shark flaunting pack- suchlike geste , successfully attacking and killing a live chick 7 m(23 ft) humpback Goliath. The shark employed the classic attack strategy used on pinnipeds when attacking the Goliath, indeed exercising the bite- and-spit tactic they employ on lower prey particulars. The Goliath was an entangled existent, heavily wasted and therefore more vulnerable to the shark' attacks. The incident is the first given attestation of great whites laboriously killing a large baleen Goliath.(128)(129) A alternate incident regarding great white shark killing humpback jumbos involving a single large lady great white nicknamed Helen was proved off the seacoast of South Africa. Working alone, the shark attacked a 33 ft(10 m) wasted and entangled humpback Goliath by attacking the Goliath's tail to cripple it before she managed to drown the Goliath by smelling onto its head and pulling it aquatic. The attack was witnessed via upstanding drone by marine biologist Ryan Johnson, who said the attack went on for roughly 50 minutes before the shark successfully killed the Goliath. Johnson suggested that the shark may have strategized its attack in order to kill such a large beast.(130)(131) Stomach contents of great whites also indicates that Goliath shark both juvenile and adult may also be included on the beast's menu, though whether this is active stalking or scavenging isn't known at present.(132)(133) Reproduction Great white shark were preliminarily allowed to reach sexual maturity at around 15 times of age, but are now believed to take far longer; manly great white shark reach sexual maturity at age 26, while females take 33 times to reach sexual maturity.(9)(134)(135) Maximum life

span was firstly believed to be further than 30 times, but a study by the Woods Hole Seaographic Institution placed it at overhead of 70 times. Examinations of vertebral growth ring count gave a maximum manly age of 73 times and a maximum womanish age of 40 times for the samples studied. The shark's late sexual maturity, low reproductive rate, long gravidity period of 11 months and slow growth make it vulnerable to pressures similar as overfishing and environmental change.(8) Little is known about the great white shark's lovemaking habits, and lovemaking geste hadn't been observed in this species until 1997 and duly proved in 2020. It was assumed preliminarily to be possible that Goliath cadavers are an important position for sexually mature shark to meet for mating.(126) According to the evidence of fisher Dick Ledgerwood, who observed two great white shark sleeping in the area near Port Chalmers and Otago Harbor, in New Zealand, it's theorized that great white shark copulate in shallow water down from feeding areas and continually roll belly to belly during coition.(136) Birth has noway been observed, but pregnant females have been examined. Great white shark are ovoviviparous, which means eggs develop and door in the uterus and continue to develop until birth.(137) The great white has an 11- month gravidity period. The shark doggy 's important jaws begin to develop in the first month. The future shark share in oophagy, in which they feed on ova produced by the mama . Delivery is in spring and summer.(138) The largest number of pups recorded for this species is 14 pups from a single mother measuring4.5 m(15 ft) that was killed apropos off Taiwan in 2019.(139)

Breaching behaviour

Great white Shark violating near Gansbaai in South Africa A breach is the result of a high- speed approach to the face with the performing momentum taking the shark incompletely or fully clear of the water. This is a stalking fashion employed by great white shark whilst hunting seals. This fashion is frequently used on cape fur seals at Seal Island in False Bay, South Africa. Because the geste is changeable, it's veritably hard to validate. It was first mugged by Chris Fallows and Rob Lawrence who developed the fashion of hauling a slow- moving seal bait to trick the shark to breach.(140) Between April and September, scientists may observe around 600 breaches. The seals swim on the face and the great white shark launch their raptorial attack from the deeper water below. They can reach pets of over to 40 km/ h(25 mph) and can at times launch themselves further than 3 m(10 ft) into the air. Just under half of observed breach attacks are successful.(141) In 2011, a 3- m-long shark jumped onto a seven- person exploration vessel off Seal Island in Mossel Bay. The crew were bearing a population study using sardines as bait, and the incident was judged not to be an attack on the boat but an accident.(142)

Natural threats

Comparison of the size of an average orca and an average great white shark Interspecific competition between the great white shark and the orca is probable in regions where salutary preferences of both species may lap.(122) An incident was proved on 4 October 1997, in the Farallon islets off California in the United States. An estimated4.7 –5.3 m(15 – 17 ft) womanish orca paralyzed an estimated 3 – 4 m(9.8 –13.1 ft) great white shark.(143) The orca held the shark upside down to induce alcohol immobility and kept the shark still for fifteen minutes, causing it to suffocate. The orca also progressed to eat the dead shark's liver.(122)(143)(144) It's believed that the scent of the taken shark's corpse caused all the great whites in the region to flee, losing an occasion for a great seasonal feed.(145) Another analogous attack supposedly passed there in 2000, but its outcome isn't clear.(146) After both attacks, the original population of about 100 great whites dissolved.(144)(146) Following the 2000 incident, a great white with a satellite tag was set up to have incontinently submerged to a depth of 500 m(1,600 ft) and swum to Hawaii.(146) In 2015, a pod of orcas was recorded to have killed a great white shark off South Australia.(147) In 2017, three great whites were set up washed ashore near Gansbaai, South Africa, with their body cavities torn open and the livers removed by what's likely to have been orcas.(148) Orcas also generally impact great white distribution. Studies published in 2019 of orca and great white shark distribution and relations around the Farallon Islands indicate that the cetaceans impact the shark negatively, with brief appearances by orcas causing the shark to seek out new feeding areas until the coming season.(149) It's unclear whether this is an illustration of competitive rejection or ecology of fear. sometimes, still, some great whites have been seen to swim near orcas without fear.(150) Attacks According to the International Shark Attack File(ISAF), between 1958 and 2016 there were 2,785 verified unprovoked shark attacks around the world, of which 439 were fatal.(15) Between 2001 and 2010, an normal of4.3 people a time failed as a result of shark attacks.(3) In 2000, there were 79 shark attacks reported worldwide, 11 of them fatal.(16) In 2005 and 2006, this number dropped to 61 and 62 independently, while the number of losses dropped to only four per time.(16) The 2016 monthly aggregate of 81 shark attacks worldwide was on par with the most recent five- time(2011 – 2015) normal of 82 incidents annually.(17) By discrepancy, the 98 shark attacks in 2015, was the loftiest monthly aggregate on record.(17) There were four losses worldwide in 2016, which is lower than the normal of eight losses per time worldwide in the 2011 – 2015 period and six deaths per annum over the once decade.(17) In 2016 58 of attacks were on surfers.(17) Despite these reports, still, the factual number of fatal shark attacks worldwide remains uncertain. For the majority of Third World littoral nations, there exists no system of reporting suspected shark attacks; thus, losses and losses near- reinforcement or at sea frequently remain unsolved or unpublicized. Of these attacks, the maturity passed in the United States(53 in 2000, 40 in 2005, and 39 in 2006).(18) The New York Times reported in July 2008 that there had been only one fatal attack in the former time.(19) On average, there are 16 shark attacks per time in the United States, with one casualty every two times.(20) According to the ISAF, the US states in which the most attacks have passed are Florida, Hawaii, California, Texas and the Carolinas, though attacks have passed in nearly every littoral state.(21) Australia has the loftiest number of fatal shark attacks in the world, with Western Australia lately getting the deadliest place in the world for shark attacks(22) with total and fatal shark mouthfuls growing exponentially over the last 40 times.(23) Since 2000 there have been 17 fatal shark attacks along the West Australian seacoast(24) with divers now facing odds of one in 16,000 for a fatal shark bite.(23)(25) Other shark attack hotspots include Réunion Island,(26) Boa Viagem in Brazil, Makena Beach, Maui, Hawaii and

Second Beach, Port St. Johns, South Africa.(27) South Africa has a high number of shark attacks along with a high casualty rate of 27 percent.(28) As of 28 June 1992,(29) Recife in Brazil began officially registering shark attacks on its beaches(substantially on the sand of Boa Viagem). Over further than two decades, 62 victims were attacked, of which 24 failed. The last deadly attack passed on 22 July 2013.(30) The attacks were caused by the species bull shark and tiger shark.(31) The shark attacks in Recife have an surprisingly high casualty rate of about 37. This is much advanced than the worldwide shark attack casualty rate, which is presently about 16, according to Florida State Museum of Natural History.(32) Several factors have contributed to the surprisingly high attack and casualty rates, including pollution from sewage runoff(33) and a(now unrestricted) original slaughterhouse.(34) The position with the most recorded shark attacks is New Smyrna Beach, Florida.(35) Developed nations similar as the United States, Australia and, to some extent, South Africa, grease further thorough attestation of shark attacks on humans than developing littoral nations. The increased use of technology has enabled Australia and the United States to record further data than other nations, which could kindly poison the results recorded. In addition to this, individuals and institutions in South Africa, the US and Australia keep a train which is regularly streamlined by an entire exploration platoon, the International Shark Attack train, and the Australian Shark Attack File. The Florida Museum of Natural History compares these statistics with the important advanced rate of deaths from other causes. For illustration, an normal of further than 38 people die annually from lightning strikes in littoral states, while lower than 1 person per year is killed by a shark in

Florida.(36)(37) In the United States, indeed considering only people who go to beaches, a person's chance of getting attacked by a shark is 1 in11.5 million, and a person's chance of getting killed by a shark is lower than 1 in264.1 million. still, in certain situations the threat of a shark attack is advanced. For illustration, in the southwest of Western Australia the chances of a surfer suffering a fatal shark bite in winter or spring are 1 in 40,000 and for divers it's 1 in 16,000.(23)(25) In comparison to the threat of a serious or fatal cycling accident, this represents 3 times the threat for a surfer and 7 times the threat for a diver.(23) Only a many species of shark are dangerous to humans. Out of further than 480 shark species, only three are responsible for two- number figures of fatal unprovoked attacks on humans the great white, barracuda and bull;(6) still, the white tip has presumably killed numerous further lepers which haven't been recorded in the statistics.(7) These shark, being large, important predators, may occasionally attack and kill people, notwithstanding the fact that all have been mugged in open water by vulnerable divers.(38)(39) The 2010 French film abysses shows footage of humans swimming coming to shark in the sea. It's possible that the shark are suitable to smell the presence of unnatural rudiments on or about the divers, similar as polyurethane diving suits and air tanks, which may lead them to accept temporary outsiders as further of a curiosity than prey. Uncostumed humans, still, similar as those surfboarding, light snorkeling or swimming, present a much lesser area of exposed skin face to shark. In addition, the presence of indeed small traces of blood, recent minor abrasions, cuts, scrapes or bruises, may lead a shark to attack a mortal in their environment. shark seek out prey through electroreception, seeing the electric fields that are generated by all creatures due to the exertion of their nerves and muscles. utmost of the oceanic white tip shark's attacks haven't been recorded,(7) unlike the other three species mentioned over. Famed naval experimenter Jacques Cousteau described the oceanic white tip as" the most dangerous of all shark".(40) ultramodern- day statistics show the oceanic white tip shark as infrequently being involved in unprovoked attacks. still, there have been a number of attacks involving this species, particularly during World War I and World War II. The oceanic white tip lives in the open sea and infrequently shows up near coasts, where utmost recorded incidents do. During the world wars, numerous boat and aircraft disasters happed in the open sea, and because of its former cornucopia, the oceanic white tip was frequently the first species on point when such a disaster happed. A case of oceanic white tip attacks include the sinking of the Nova Scotia, a British steamship carrying 1,000 people that was torpedoed by a German submarine on 18 November 1942, near South Africa. Only 192 people survived, with numerous deaths attributed to the oceanic white tip shark.(41) The same species is believed to have been responsible for numerous of the 600 – 800 or further casualties following the torpedoing of the USS Indianapolis on 30 July 1945.(42) Black December refers to at least nine shark attacks on humans, causing six deaths, that passed along the seacoast of KwaZulu- Natal Province, South Africa, from 18 December 1957, to 5 April 1958.(43) In addition to the four species responsible for a significant number of fatal attacks on humans, a number of other species have attacked humans without being provoked, and have on extremely rare occasions been responsible for a mortal death. This group includes the shortfin Mako, hammerhead, Galapagos, slate reef, blacktip, lemon, silky shark and blue shark.(6) These shark are also large, important predators which can be provoked simply by being in the water at the wrong time and place, but they're typically considered less dangerous to humans than the former group. On the evening of 16 March 2009, a new addition was made to the list of shark known to have attacked mortal beings. In a painful but not directly life- hanging incident, a long- distance swimmer crossing the Alenuihaha Channel between the islands of Hawai and Maui was attacked by a cookiecutter shark. The two mouthfuls were delivered about 15 seconds piecemeal.(44)

The three most commonly involved sharks

- The <u>great white shark</u> is involved in the most fatal unprovoked attacks[45]

- The <u>tiger shark</u> ranks as the second most fatal in unprovoked attacks[45]

- The <u>bull shark</u> ranks as the third most fatal in unprovoked attacks[45]

Types of attacks

Shark attack indicators use different criteria to determine if an attack was" provoked" or" unprovoked." When considered from the shark's point of view, attacks on humans who are perceived as a hazard to the shark or a contender to its food source are each" provoked" attacks. Neither the International Shark Attack File(ISAF) nor the Global shark Attack File(GSAF) accord casualties of air/

sea disasters" provoked" or" unprovoked" status; these incidents are considered to be a separate kind.(46)(47) Posthumous scavenging of mortal remains(generally drowning victims) are also not accorded" provoked" or" unprovoked" status.(47)(45) The GSAF categorizes scavenging mouthfuls on humans as" questionable incidents."(47) The most common criteria for determining" provoked" and" unprovoked" attacks are bandied below Provoked attack Provoked attacks do when a mortal touches, hooks, nets, or else aggravates the animal. Incidents that do outside of a shark's natural niche, similar as aquariums and exploration holding- pens, are considered provoked, as are all incidents involving captured shark. occasionally humans inadvertently provoke an attack, similar as when a surfer accidentally hits a shark with a surf board. Unprovoked attack Unprovoked attacks are initiated by the shark — they do in a shark's natural territory on a live human and without mortal provocation.(46)(47) There are three subcategories of unprovoked attack • Hit- and- run attack – generallynon-fatal, the shark mouthfuls and also leaves; most victims don't see the shark. This is the most common type of attack and generally occurs in the surf zone or in murky water. utmost megahit- and- run attacks are believed to be the result of incorrect identity.(48) • Sneak attack – the victim won't generally see the shark, and may sustain multiple deep mouthfuls. This kind of attack is raptorial in nature and is frequently carried out with the intention of consuming the victim. It's extraordinarily rare for this to do. • Bump- and- suck attack – the shark circles and bumps the victim before smelling. Great whites are known to do this on occasion, appertained to as a" test bite", in which the great white is trying to identify what's being stunk. Repeated mouthfuls, depending on the response of the victim(thrashing or scarifying may lead the shark to believe the victim is prey), aren't uncommon and can be severe or fatal. Bump- and-suck attacks aren't believed to be the result of incorrect identity.(48) An incident passed in 2011 when a 3- metre long(500 kg) great white shark jumped onto a 7- person exploration vessel off Seal Island, South Africa. The crew were bearing a population study using sardines as bait and originally retreated to safety in the arc of the boat while the shark thrashed about, damaging stuff and energy lines. To keep the shark alive while a rescue boat hauled the exploration vessel to shore, the crew poured water over its gills and ultimately used a pump for mechanical ventilation.

The shark was eventually lifted back into the water by crane and, after getting disoriented and beaching itself in the harbor, was successfully hauled out to sea, swimming down. The incident was judged to be an accident.(49) Reasons for attacks Large shark species are apex predators in their terrain,(50) and therefore have little fear of any critter(other than orcas(51)) with which they cross paths. Like utmost sophisticated nimrods, they're curious when they encounter commodity unusual in their homes. Lacking any branches with sensitive integers similar as hands or bases, the only way they can explore an object or organism is to suck it; these mouthfuls are known as test mouthfuls.(52) Generally, shark mouthfuls are exploratory, and the beast will swim down after one bite.(52) For illustration, exploratory mouthfuls on surfers are allowed to be caused by the shark mistaking the surfer and surfboard for the shape of prey.(53) nevertheless, a single bite can dolorously injure a mortal if the beast involved is a important bloodsucker similar as a great white or barracuda shark.(54) A shark will typically make one nippy attack and also retreat to stay for the victim to die or weaken from shock and blood loss, before returning to feed. This protects the shark from injury from a wounded and aggressive target; still, it also allows humans time to get out of the water and survive.(55) shark attacks may also do due to territorial reasons or as dominance over another shark species, performing in an attack.(56) shark are equipped with sensitive organs called the Ampullae of Lorenzini that descry the electricity generated by muscle movement.(57) The shark's electrical receptors, which pick up movement, descry signals like those emitted from fish wounded, for illustration, by someone who's spearfishing, leading the shark to attack the person by mistake.(56) GeorgeH. Burgess, director of the International Shark Attack File, said the following regarding why people are attacked" Attacks are principally an odds game grounded on how numerous hours you're in the water".(58)

CHAPTER THREE

The Theory of Evolution of Megalodons

The Megalodons suffered extinction due to the scarcity of food during the start of the ice age. As the ice age started to take over the Megalodons were cornered away from the warmer waters where the whales and other species in its food chain had already migrated. Till today Scientists have an ongoing debate on whether the Megalodon did or did not evolve into the living day Great whites. Most people believe that the Megalodons did not evolve into anything and that they went extinct completely, while some believe that they have evolved into the living day Mako sharks which are the fastest species of sharks.

There are more than 470 species of sharks in this world, even more if we include the prehistoric species. Now Scientists state that the Megalodons have died due to starvation as the animals that were present in the Megalodon's food chain had migrated to warmer water and the Megalodons were not able to reach the warm waters before the ice age progressed. A study states that after the

Megalodons getting trapped a new species emerged and that species hunted and fed on the Megalodons and this species was The Great whites. The study states that the great whites came into existence as a completely different species. The Megalodons lurked Japanese water while on the other hand the great whites prefer to lurk in cold waters.

Now I would like to say that this theory is partly correct by my point of view. But, I have a different theory to propose.

This theory goes in such a way that if the Megalodons went extinct because of starvation caused by the scarcity for which the ice age was responsible, so they must have faced malnutrition at some point in their life after the ice age. [C]. Malnutrition is when an organism doesn't get enough nutrition for the growth of its body. And so the organism turns out with an unusual body structure, malnutrition can also affect the functioning of the brain, if the brain does not get enough nutrients to grow it ends up with a bad growth rate and in humans it causes bad memory, speech or coordination. It can also cause dwarfism. The megalodons should have faced the same problems in the condition where the food was limited. As normal organisms they must have suffered malnutrition. The size of Megalodons was between 30-60ft, [D], but due to the malnutrition the infants of a healthy Megalodon must have grown a bit smaller than the normal size. As the scarcity of food kept on the infants must have started to grow much smaller sizes. Slowly but surely they must have grown to the size of a modern day Great white around 15-20ft. According to an article published by allthatisinteresting.com, the infant of the Megalodon was nearly about 6ft. [B]. So if the Infant was naturally 6ft,

Because of mal-nourishment it would have been 4-5ft or lesser.

There is a color difference between the megalodons and great whites, well I happen to have an explanation on the color change; the great whites have black colored skin unlike, like the megalodons who are believed to have a grayish brown skin color. This change could be of the ice age. When the sharks must have faced extreme cold environment, their skin color would have changed to black as the color black is a good absorber of heat, the sharks must have used it to absorb as much heat possible from the sun. The new generation of megalodons must have used this technique to survive the cold waters and when the new generation got hungry they must have fed on the earlier generation and even that could be included in the causes of the extinction of the Megalodon.

The Great whites are considered to be much more aggressive than Megalodon with 2785 human attacks recorded from the year 1958 to 2016 from which 439 were fatal. A Megalodon is considered to have a prey as much as its own size like the blue whale and even if it existed in today's date it wouldn't have attacked humans as the great white because we would have been too small for it to even fill his hunger, while on the other hand the great white is seen attacking humans. The great white, on the other hand, is known to be an aggressive predator and has an extremely muscular body, capable of chasing down some of the fastest swimmers in the sea. Well any living organisms which can defend themselves do not attack unless provoked. According to a study there were 57 unprovoked attacks world wide of which 10 were fatal. But there is no doubt that the Great white is indeed an aggressive predator.

The teeth of sharks keep getting replaced, Sharks do not rely on two sets of teeth – they have an endless supply of teeth, with a dentition that regenerates constantly throughout life. In some sharks, a new set of teeth develops every two weeks! Sharks don't actually re-grow teeth one by one but have multiple rows inside their jaw that are constantly regrown. When a tooth on the edge of the jaw drops out, the corresponding tooth in the row behind it moves forward to replace it. The underlying soft tissues anchor can carry each tooth like a conveyor belt. The teeth of a megalodon measure 7 inches and the teeth of the Great whites are about 6.6 inches a difference of 6 inches. This could be explained by the malnutrition situation caused in the ice age. Due to the scarcity of food the megalodons would also be short on vitamins and due to the deficiency of calcium their teeth could have shrunk as they didn't get a good supply of calcium. Therefore the teeth must have shrunk from 7inches to 6.6 inches.

Now all of these changes did not happen overnight, they happened to each different generation. As one generation gave birth to another, the changes took place in each generation. First was the malnutrition the came along the change in size and later the change in diet followed by change in behavior, color and then lastly territories.

<u>Charles Darwin's Theory of Evolution states: gradual changes occurring on organisms over a long duration is called evolution. Organisms must show certain modification in order to better adapt to their surroundings. The ones who don't perish, while the ones who do, grow and reproduce.</u>

Taking account to this theory, the change in size, diet, behavior, color, territories we should consider that megalodons evolved in to the Modern day's Great white, but we should not consider that this species cannot return to its earlier state, the largest Great white ever recorded till today was about 20ft in length and is now 50 years old and is expected to live for another 20 years. If there is enough food for them to consume they will continue to grow up to their fullest size. And maybe one day they may out grow the normal growth rate of a Great white, who knows maybe we can see them evolving backwards and maybe they could grow around the size of the Titan creature that once lurked as the largest in the sea, Just like the scientists believe that the modern day crocodiles which are the descendants of the Sarcosuchus, may evolve back into them as the largest Crocodile recorded was about 20ft and his name was Lolong, he passed away back in 2013.

CHAPTER FOUR

Bringing them back to life

The extinct species of shark can be brought back but I would prefer not to do so. These creatures went extinct by a natural cause. An event caused by nature made them go extinct. Sharks are neither mammals nor Oviparous they produce eggs but don't lay any, the infants are born within the mother and later released into the water by the mother. A life of a shark is hard as at time of hatching of eggs, the one who has hatched first starts eating its siblings, a perfect example of "The survival of the Fittest".

Scientist still research on how this species could be made to come back to the living. Fortunately there seems to be a solution. We have discovered how an animal could be cloned for example: A dog can be cloned by taking a sample of its DNA which could be obtained by collecting the tissue samples of the dog you wish to clone and then separating the tissue from the cells. Then the cells containing the DNA are put in a centrifuge machine, which allows the scientist to obtain the DNA. Then the eggs of another dog are emptied, meaning: all the DNA present in the eggs are removed. Then the genetic empty eggs are injected with the DNA obtained for the cells and later the eggs of the dog grow into embryos by the electro cell manipulator, which are then transferred to a surrogate. Then later the surrogate gives birth to these genetically cloned puppies.

The same method could be used in the case of sharks. <u>Even if we get the tiniest sample of the tissue of a megalodon, we can extract the DNA from the tissue.</u> Which seems Impossible as cartilage is one of the softest bones present, it decays pretty easily but the teeth dug out could become a major source of DNA extraction as the hard covering must have protected the DNA from decaying or getting contaminated. [A]

We have been able to dig out megalodon teeth. But once they are put in a museum they are of no use, what if we extract the DNA present in the teeth?

Incubation

The megalodon would be way too big to be transferred in a surrogate. If we do that we risk killing the surrogate. So we must find a way to fertilize the eggs outside the mother's womb. We can use the sperm of another shark by doing the same method of extracting the DNA from the sperm and injecting the DNA of the megalodon obtained from the tissue sample. This way we have a genetically designed sperm which can now be used to fertilize the shark eggs.

The eggs could be transferred to an incubator or perhaps a containment unit made for the fetus to grow. The Amniotic fluid is a combination of proteins, nutrients, hormones, antibodies and the urine of the fetus. Amniotic fluid the liquid that surrounds the fetus in the womb during pregnancy could be formed by first water with proteins and nutrients, then later the embryo will release hormones as the fetus starts to grow after a certain period of time the fetus will form kidneys which will process the urine which will help in forming the amniotic fluid. Now even after the embryo is in a man-made containment unit there should be a way where we could somehow fill the incubator with amniotic fluid. Now as the womb is a man made unit, in normal cases whatever the mother consumes provides essential nutrients to the fetus. But in this case we will have to supply nutrients and proteins through the form of liquids/fluids. The embryo with utilize the nutrients and proteins from the fluid around it. As the embryo grows bones and organs we can start feeding the fetus nutrients necessary for the growth by injecting the nutrients in its body. At time of growth it will be growing sizes larger than our incubator. For that we must be ready with incubators with various sizes, so as it outgrows the first incubator we can transfer it to the next one or make one big enough to contain it until its body starts functioning properly.

CHAPTER SIX

EFDU- External Fetal Development Unit

The perimeter of the incubator would be a sturdy cylindrical sheet of transparent polycarbonate. There would be a set of thermometers inside the incubators to measure and maintain the temperature of the incubator, so that the fetus growing inside will develop. Once the fetus develops an umbilical cord the nutrients would be supplied through a synthesized umbilical cord. The design was of a flexible pipe in which there would be 3 outlets: 2 veins and an artery. This will perform their respective tasks. If we could pass the pipe through the umbilical cord of the fetus by somehow separating the placenta and connecting each vein and artery, we can supply the necessary nutrition to the fetus in order to continue its growth. The nutrients would obviously be in a liquid form. where on the other hand the blood supply would be of a Shark or the same species; we cannot obtain Megalodon blood but what if we could alter the DNA of a shark with similar characteristics. We are limited by the technology of today's world, but who knows technology could improve in the next decade itself. When and if we are able to synthesize blood, it would be a real revolutionary step towards the field of Genetic engineering. However if we do synthesize blood at some point in time, that blood could be used in the supply of oxygen to the fetus. The blood will of-course be processed by oxygenating it etc, and then let in the infant's body.

In the first embryonic stage of the shark, it will continue to grow through mitosis by dividing cells. The moment where it starts to form organs such as brain, kidneys, or bone structure (cartilage), it'll be in a desperate need of nutrients that's when we take part in the whole process. The connecting of umbilical cord is a very crucial activity of the entire process, one mistake may lead to an infinite amount of cascades and the whole project will be compromised, not only causing loss of money that has been put in this process but also loss of an endangered life. So the umbilical cord will be connected only by professionals and no other is to be let near it. Once the cord is attached our one and only job would be to make sure that the fetus get's the necessary nutrients i.e our synthesized cord works. Until the shark reaches to a stage where it can swim around it is to be supplied with the cord only. Once it gains the ability to swim it can be fed through the medium of whole pieces of fish.

There will be a crew of scientists including:

1. Biologists
2. Embryologists
3. Technicians. Etc

Each having a staring hand in their area of expertise will monitor the growth/development of the fetus. Also, the technicians will check the equipments every single day, in case of a malfunction a team would be present to fix the malfunction. Reviving an extinct species may or may not be a good idea but think about the unlimited applications of this incubator. It could be used as a surrogate for human infants as well. Women who cannot hold pregnancy can externally fertilize their eggs which would be developed in to embryos by the fertilization process of sperm originating from their own respective partners. These embryos formed outside the mother womb will be released in the incubator for further development, only in this case we will not need a synthesized source of blood because the source would be of the mothers blood type. However if the blood is not accepted by the fetus we will have to revert back to synthesizing blood with respect to the same genetic content.

I agree that all of the ideas/theories mentioned in this paper of mine are altogether a giant hypothesis, but I will work hard to transform this idea of mine into a real life project in the further years of my life.

References

https://www.oneindia.com/2008/08/12/dna-sharks-teeth-protect-endangered-species-1218535920.html - [A]

https://allthatsinteresting.com/megalodon-cannibal-study - [B]

https://www.livescience.com/63361-megalodon-facts.html - [C]

https://www.britannica.com/animal/megalodon - [D]

https://en.wikipedia.org/wiki/Great_white_shark - [E]

1. *"Yearly Worldwide Shark Attack Summary"*. International Shark Attack File. Florida Museum of Natural History, University of Florida. *Retrieved 27 November 2017*. The 2016 yearly total of 81 unprovoked attacks was on par with our most recent five-year (2011–2015) average of 82 incidents annually.

2. ^*"ASU shark scientist: Fatal shark attacks 'extremely rare'"*. ASU News. 6 August 2020.

3. ^ Jump up to:*ab"Shark attacks are rare – and related deaths even rarer"*. The Guardian. 17 August 2011.

4. ^ Plumer, Brad (8 July 2014). *"How common are shark attacks, really?"*. Vox.

5. ^*"Chart: The animals that are most likely to kill you this summer - The Washington Post"*.

6. ^ Jump up to:*abc* ISAF Statistics on Attacking Species of Shark

7. ^ Jump up to:*abc* Edmonds, Molly (5 June 2008). *"HowStuffWorks "Dangerous Shark 4: Oceanic White*

tip Shark"". Animals.howstuffworks.com. *Retrieved 23 September 2010*.

8. ^ Pepin-Neff, Christopher; Hueter, Robert (23 January 2013). *"Science, policy, and the public discourse of shark "attack": a proposal for reclassifying human–shark interactions"*. Journal of Environmental Studies and Sciences. **3** (1): 65–73. *doi*:*10.1007/s13412-013-0107-2*.

9. ^*"Total shark attacks per year"*. Our World in Data.

10. ^ Jump up to:[ab]*"Incidents List"*. *Government of Hawaii*. 18 June 2014. *Retrieved 27 May 2019*.

11. ^ Lam, Kristin (27 May 2019). *The Associated Press* (ed.). *"California man, 65, swimming off Maui killed in Hawaii's first fatal shark attack since 2015"*. *USA Today*. *Gannett. Retrieved 27 May 2019*.

12. ^*"A Teen Died After Losing His Leg and Penis in a Horrific Shark Attack"*. 4 June 2018.

13. ^*"Confirmed Unprovoked Shark Attacks (1847–Present). Europe"*. *Florida Museum of Natural History*. 24 January 2018. *Retrieved 27 May 2019*.

14. ^*"Un surfeur tué après une attaque de requin à La Réunion"*. BFMTV. 9 May 2019.

15. ^*"World's Confirmed Unprovoked Shark Attacks"*. International Shark Attack File. 25 August 2015. *Retrieved 10 September 2015*.

16. ^ Jump up to:[ab]*"ISAF Statistics for the Top Ten Worldwide Locations with the Highest Shark Attack Activity (1999–2009)"*. *Florida Museum of Natural History* Flmnh.ufl.edu. 25 March 2010. *Retrieved 23 September 2010*.

17. ^ Jump up to:[abcd]*"ISAF 2016 Worldwide Shark Attack Summary"*. Florida Museum of Natural History. *Retrieved 24 April 2017*.

18. ^*"ISAF Statistics for the USA Locations with the Highest Shark Attack Activity Since 1999"*. Flmnh.ufl.edu. 3 May 2010. *Retrieved 23 September 2010*.

19. ^ Tierney, John (29 July 2008). *"10 Things to Scratch From Your Worry List"*. The New York Times. *Retrieved 19 September 2010*.

20. ^*"Shark Facts: Attack Stats, Record Swims, More"*. News.nationalgeographic.com. 28 October 2010. *Retrieved 17 August 2012*.

21. ^*"Map of United States (incl. Hawaii) Confirmed Unprovoked Shark Attacks"*. Flmnh.ufl.edu. 26 August 2010. *Retrieved 16 February 2012*.

22. ^*"WA 'deadiest' for shark attacks – The West Australian"*. Au.news.yahoo.com. 1 April 2012. *Retrieved 17 August 2012*.

23. ^ Jump up to:[abcd] Sprivulis, P (2014). *"Western Australian Coastal Shark Bites: A risk assessment"*. Australasian Medical Journal. **7** (2): 137–42. *doi*:*10.4066/AMJ.2014.2008*. *PMC3941575*. *PMID24611078*.

24. ^*"Timeline of shark attacks along the Western Australian coast"*. WA Today. *Retrieved 11 October 2020*.

25. ^ Jump up to:[ab]*"Shark attacks and whale migration in Western Australia"*. *Retrieved 25 April 2017*.

26. ^ Ballas, Richard; Saetta, Ghislain; Peuchot, Charline; Elkienbaum, Philippe; Poinsot, Emmanuelle (2017). *"Clinical features of 27 shark attack cases on La Réunion Island (PDF Download Available)"*. Journal of Trauma and Acute Care Surgery. **82** (5): 952–955. *doi*:*10.1097/TA.0000000000001399*. *PMID28248805*. *S2CID21996541. Retrieved 23 July 2017*.

27. ^*"The World's 10 Deadliest Shark Attack Beaches"*. The

Inertia. *Retrieved 25 April 2017*.

28. ^*"Map of World's Confirmed Unprovoked Shark Attacks"*. Flmnh.ufl.edu. 6 January 2011. *Retrieved 25 November 2011*.

29. ^*"Shark Attack Data Brazil"*. Shark Attack Data. *Retrieved 23 December 2017.*

30. ^*"Tubarão: julho é o mês com mais ataques. Saiba como evitá-los"*. Jornal do Comércio. 18 July 2016. *Retrieved 23 December 2017*.

31. ^*"Ataques de tubarão em Recife – Conheça a verdade e as causas desse fenômeno"*. Visitar Recife. 27 November 2015. *Retrieved 23 December 2017*.

32. ^*"The beautiful Brazilian beaches plagued by shark attacks"*. BBC. 27 September 2012. *Retrieved 23 December 2017*.

33. ^*"The beautiful Brazilian beaches plagued by shark attacks"*. BBC. 27 September 2012. *Retrieved 6 December 2019*.

34. ^*"SHARK WEEK! FLORIDA EXPERT WADES INTO SCARY BRAZIL WATERS"*. South Florida Sun-Sentinel. 29 July 2006. *Retrieved 6 December 2019*.

35. ^ Regenold, Stephen (21 April 2008). *"North America's top shark-attack beaches"*. USA Today. *Retrieved 9 April 2010*.

36. ^*"The Relative Risk of Shark Attacks to Humans"*. Flmnh.ufl.edu. *Retrieved 23 September 2010*.

37. ^ Florida Museum of Natural History, University of Florida. **A Comparison with the Number of Lightning Fatalities in Coastal United States: 1959–2006**

38. ^*"Hawaiian newspaper article"*. Honoluluadvertiser.com. *Retrieved 23 September 2010*.

39. ^ The **1992 Cageless shark-diving expedition** by Ron and Valerie Taylor.

40. ^ Cousteau, Jacques-Yves & Cousteau, Philippe (1970). The Shark: Splendid Savage of the Sea. Doubleday & Company, Inc.

41. ^ Bass, A.J., J.D. D'Aubrey & N. Kistnasamy. 1973. "Sharks of the east coast of southern Africa. 1. The genus Carcharhinus (Carcharhinidae)." Invest. Rep. Seaogr. Res. Inst., Durban, no. 33, 168 pp.

42. ^ Martin, R. Aidan. *"Elasmo Research"*. ReefQuest. *Retrieved 6 February 2006*.

43. ^*"South Africa Rethinks Use of Shark Nets"*. *Retrieved 22 April 2017*.

44. ^ University of Florida News New study documents first cookiecutter shark attack on a live human

45. ^ Jump up to:[abcd]*"ISAF Statistics on Attacking Species of Shark"*. Global Shark Attack File. 9 May 2020. *Retrieved 9 May 2020*.

46. ^ Jump up to:[ab]*Burgess, George H."ISAF 2011 Worldwide Shark Attack Summary"*. Global Shark Attack File. *Retrieved 26 June 2012*.

47. ^ Jump up to:[abcd]*"Incident Log"*. Global Shark Attack File. *Retrieved 26 June 2012*.

48. ^ Jump up to:[ab]*Burgess, George H."How, When, & Where Sharks Attack"*. International Shark Attack File. *Retrieved 26 June 2012*.

49. ^ Rice, Xan (19 July 2011). *"Great white shark jumps from sea into research boat"*. *The Guardian*. London. *Retrieved 20 July 2011*. Marine researchers in South Africa had a narrow escape after a three-metre-longGreat white breached the surface of the sea and leaped into their boat, becoming trapped on deck for more than an hour. [...] Enrico Gennari, an expert onGreat whites, [...] said it was almost certainly an accident rather than an attack on the boat.

50. ^*"Apex Predators Program"*. Na.nefsc.noaa.gov. Archived from *the original* on 10 February 2014. *Retrieved* 23 September *2010*.

51. ^ Turner, Pamela S. (October–November 2004). *"Showdown at Sea: What happens whenGreat whites go fin-to-fin with killer whales?"*. National Wildlife. **42** (6). *Retrieved* 21 August *2012*.

52. *"What To Expect on YourGreat white Diving Tour"*. Romow.com. 7 August 2009. *Retrieved* 23 September *2010*.

53. ^ Grabianowski, Ed (10 August 2005). *"HowStuffWorks "How Shark Attacks Work""*. Adventure.howstuffworks.com. *Retrieved* 23 September *2010*.

54. ^ Grabianowski, Ed (10 August 2005). *"HowStuffWorks "Shark Attack Damage""*. Adventure.howstuffworks.com. *Retrieved* 23 September *2010*.

55. ^*"Great White Shark"*. Extremescience.com. *Retrieved* 23 September *2010*.

56. ^ Jump up to:[ab] Grabianowski, Ed (10 August 2005). *"HowStuffWorks "Shark Sensory System""*. Adventure.howstuffworks.com. *Retrieved* 23 September *2010*.

57. ^*"Ampullae of Lorenzini"*. Marinebiodiversity.ca. *Retrieved* 23 September *2010*.

58. ^*"Shark attacks at record high"*. BBC News. 9 February 2001. *Retrieved* 6 April *2010*.

59. ^ Meerman, Ruben (16 January 2009). *"Shark nets"*. ABC Science. *Archived* from the original on 29 November 2016. *Retrieved* 5 January *2017*.

60. ^*"Fact File: Protecting people from shark attacks"*. ABC News. 5 January 2015.

61. ^ Jump up to:*ab* Carroll Nick. *"Nick Carroll On-Beyond the Panic, The Facts about Shark Nets"*. Coastal Watch. Archived from *the original* on 5 March 2017. *Retrieved 22 November 2016*.
62. ^ Media, Australian Community Media – Fairfax (8 November 2016). *"Beyond the panic: the facts about shark nets". Retrieved 22 April 2017.*
63. ^ Jump up to:*abcd"Shark Attacks and the Surfer's Dilemma: Cull or Conserve?"*. 21 August 2015. *Retrieved 22 April 2017*.
64. ^ Jump up to:*abc*http://www.sharkangels.org/index.php/media/news/157-shark-netsArchived 19 September 2018 at the Wayback Machine*Shark Nets*. Sharkangels.org. Retrieved 18 September 2018.
65. ^ Jump up to:*ab* Curtis; et al. (2012). *"Responding to the risk of white shark attack: updated statistics, prevention, control, methods and recommendation. Chapter 29 In: M. L. Domeier (ed). Global Perspectives on the Biology and Life History of the White Shark"*. CRC Press. Boca Raton, FL. *Retrieved 18 January 2020*.
66. ^*"Can governments protect people from killer sharks?"*. ABC News. 22 December 2013. *Retrieved 22 April 2017*.
67. ^*"Sydney shark nets set to stay despite drumline success | Swellnet Dispatch | Swellnet"*. Swellnet.com. *Sydney shark nets set to stay despite drumline success.* Bruce Mackenzie. 4 August 2018. Retrieved 20 September 2018.
68. ^*"Shark culling · Save our Sharks · Australian Marine Conservation Society"*. Archived from *the original* on 2 October 2018. Marineconservation.org.au. Shark culling (archived). Retrieved 2 December 2018.
69. ^*"NSW govt won't back down on shark nets | SBS News"*. *NSW govt won't back down on shark nets.* Sbs.com.au.

Retrieved 18 September 2018.

70. ^*"Here's What You Need To Know About Australia's SMART Drum Lines Being Used To Prevent Shark Attacks"*. *Here's What You Need To Know About Australia's SMART Drum Lines Being Used To Prevent Shark Attacks*. Elfy Scott. 5 July 2018. Retrieved 18 September 2018.

71. ^*"The Importance of Sharks | Seaa Europe"*. Eu.seaa.org. *The Importance of Sharks in the Ecosystem*. Retrieved 18 September 2018.

72. ^ Jump up to:[ab] Alana Schetzer (8 May 2017). *"Sharks: How a cull could ruin an ecosystem | Pursuit by The University of Melbourne"*. *"Sharks: How a cull could ruin an ecosystem"*. Science Matters. University of Melbourne – via Pursuit. Retrieved 18 September 2018.

73. ^ Department of Primary Industries and Fisheries (March 2006), *A Report on the Queensland Shark Safety Program(PDF)*, Queensland Government, archived from *the original(PDF)* on 23 January 2014*, retrieved* 6 January *2017*

74. ^ Dudley, S.F.J. (1997). *"A comparison of the shark control programs of New South Wales and Queensland (Australia) and KwaZulu-Natal (South Africa)"*. Sea Coast Manag. **34** (34): 1–27. *doi:10.1016/S0964-5691(96)00061-0*.

75. ^ Hazin, F. H. V.; Afonso, A. S. (1 August 2014). "A green strategy for shark attack mitigation off Recife, Brazil". Animal Conservation. **17** (4): 287–296. *doi:10.1111/acv.12096*. *hdl:10400.1/11160*.

76. ^ Dudley, Sheldon F.J.; Haestier, R.C.; Cox, K.R.; Murray, M. (January 1998). "Shark control: Experimental fishing with baited drumlines". Marine and Freshwater

Research. **49** (7): 653. _doi_:_10.1071/MF98026_.

77. ^_"NSW North Coast SMART drumline data"_. NSW Government: Department of Primary Industries. _Retrieved_ 4 December _2018_.

78. ^_"Queensland's Shark Control Program Has Snagged 84,000 Animals. Latest news"_. Action for Dolphins. _Queensland's Shark Control Program Has Snagged 84,000 Animals._ Thom Mitchell. 20 November 2015. Retrieved 25 December 2018.

79. ^_"Sharks - Marine Science Australia"_. _Sharks – Marine Science Australia._ Ausmarinescience.com. Retrieved 18 September 2018.

80. ^http://www.seashepherd.org.au/apex-harmony/overview/queensland.htmlArchived 23 August 2017 at the Wayback Machine_Queensland – Overview._ Seashepherd.org.au. Retrieved 18 September 2018.

81. ^ Calla Wahlquist (12 February 2015). _"Western Australia's 'serious threat' shark policy condemned by Senate | Western Australia"_. The Guardian. _"Western Australia's 'serious threat' shark policy condemned by Senate"._ Retrieved 18 September 2018

82. ^ Carl Meyer (11 December 2013). _"Western Australia's shark culls lack bite (and science)"_. _"Western Australia's shark culls lack bite (and science)"._ Theconversation.com. Retrieved 18 September 2018.

83. ^ Chloe Hubbard (30 April 2017). _"No Shark Cull: Why Some Surfers Don't Want to Kill Great Whites Despite Lethal Attacks"_. _"No Shark Cull: Why Some Surfers Don't Want to Kill Great Whites Despite Lethal Attacks"._ NBC News. Retrieved 18 September 2018.

84. ^ Green, M., Ganassin, C. and Reid, D. D. _"Report into the NSW Shark Meshing (Bather Protection)_

Program"(PDF). State of New South Wales through NSW Department of Primary Industries. Archived from *the original*(PDF) on 16 June 2016. *Retrieved 21 December 2016*.

85. ^ Robbins, William D.; et al. (2014). *"Experimental evaluation of shark detection rates by aerial observers"*. PLOS ONE. **9** (2): e83456. *Bibcode:2014PLoSO...983456R*. *doi:10.1371/journal.pone.0083456*. *PMC3911894*. *PMID24498258*.

86. ^*"Thousands protest over shark cull in Australia"*, The Telegraph, UK, 1 February 2014, *archived* from the original on 18 November 2021, *retrieved 5 January 2017*

87. ^*Open letter to WA Government re: Proposal to use drum lines for shark population control and targeting of sharks entering protected beach*(PDF), Support Our Sharks, 2013, *retrieved 5 January 2017*

88. ^ McCall, Matt; 02, National Geographic PUBLISHED July. *"2 Weeks, 4 Deaths, and the Beginning of America's Fear of Sharks"*. National Geographic News. *Retrieved 28 July 2015*.

89. ^attack survivors unite to save sharks, Australian Geographic, 14 September 2010

90. ^ Choi, Charles Q. *"How 'Jaws' Forever Changed Our View ofGreat whites"*. Web. Live Science. *Retrieved 29 August 2013*.

91. ^ Beryl, Francis (2011). *"Before and after Jaws: Changing representations of shark attacks"*. researchrepository.murdoch.edu.au. *Retrieved 6 December 2017*.

92. ^*"Shark Week : Discovery Channel"*. Dsc.discovery.com. *Retrieved 17 August 2012*.

93. ^*"Dangerous shores for Jaws"*. The Olive Press. Luke Stewart Media *SL*. 28 March 2008. *Retrieved 10*

December 2016.

94. ^ McGowan, Cliodhna (1 February 2007). *"ISDHF looks to permanent home"*. Caycom Pass. Archived from *the original* on 16 July 2007. *Retrieved 20 June 2009*.

95. ^ Smith, Emily (29 August 2021). *"Shark attack survivor and drumline contractor say conservation group supporters harass them"*. ABC News. *Retrieved 29 August 2021*.

Page- 7

1. Rigby, C.L.; Barreto, R.; Carlson, J.; Fernando, D.; Fordham, S.; Francis, M.P.; Herman, K.; Jabado, R.W.; Liu, K.M.; Lowe, C.G.; Marshall, A.; Pacoureau, N.; Romanov, E.; Sherley, R.B.; Winker, H. (2019). *"Carcharodon carcharias"*. *IUCN Red List of Threatened Species*. **2019**: e.T3855A2878674. *doi:10.2305/ IUCN.UK.2019-3.RLTS.T3855A2878674.en*. *Retrieved 19 November 2021*.

2. ^*"Appendices | CITES"*. cites.org. *Retrieved 14 January 2022*.

3. ^*"Great white sharks: 10 myths debunked"*. The Guardian. *Retrieved 3 June 2016*.

4. ^ Jump up to:[abcdefg] Viegas, Jennifer. *"LargestGreat white Don't Outweigh Whales, but They Hold Their Own"*. Discovery Channel. Archived from *the original* on 7 February 2010. *Retrieved 19 January 2010*.

5. ^ Jump up to:[ab]*"Just the Facts Please"*. GreatWhite.org. *Retrieved 3 June 2016*.

6. ^ Jump up to:[ab] Parrish, M. *"How Big areGreat whites?"*. Smithsonian National Museum of Natural History Sea Portal. *Retrieved 3 June 2016*.

7. ^*"Carcharodon carcharias"*. Animal Diversity Web.

Retrieved 5 June 2016.

8. ^ Jump up to:*ab"New study finds extreme longevity in white sharks"*. Science Daily. 9 January 2014.

9. ^ Jump up to:*abc* Ghose, Tia (19 February 2015). *"Great White Sharks Are Late Bloomers"*. LiveScience.com.

10. ^ Klimley, A. Peter; Le Boeuf, Burney J.; Cantara, Kelly M.; Richert, John E.; Davis, Scott F.; Van Sommeran, Sean; Kelly, John T. (19 March 2001). "The hunting strategy of white sharks (Carcharodon carcharias) near a seal colony". Marine Biology. **138** (3): 617–636. *doi*:*10.1007/s002270000489*. *ISSN0025-3162*. *S2CID85018712*.

11. ^ Jump up to:*abc* Thomas, Pete (5 April 2010). *"Great white shark amazes scientists with 4,000-foot dive into abyss"*. GrindTV. Archived from *the original* on 17 August 2012.

12. *^Currents of Contrast: Life in Southern Africa's Two Seas*. Struik. 2005. pp. 31–. *ISBN978-1-77007-086-8*.

13. ^ Knickle, Craig. *"Tiger Shark"*. Florida Museum of Natural History Ichthyology Department. Archived from *the original* on 7 July 2013. *Retrieved 2 July 2009.*

14. *^"ISAF Statistics on Attacking Species of Shark"*. *Florida Museum of Natural History* University of Florida. Archived from *the original* on 24 April 2012. *Retrieved 4 May 2008.*

15. ^ Jump up to:*ab"Carcharodon carcharias"*. UNEP-WCMC Species Database: CITES-Listed Species On the World Wide Web. Archived from *the original* on 16 June 2013. *Retrieved 8 April 2010.*

16. ^ Jump up to:*abcRecovery Plan for the White Shark (Carcharodon carcharias)* (Report). Government of Australia Department of Sustainability, Environment, Water, Population and Communities. 2013.

17. ^ Cronin, Melissa (10 January 2016). *"Here's Why We've Never Been Able to Tame theGreat white"*.

18. ^ Hile, Jennifer (23 January 2004). *"Great White Shark Attacks: Defanging the Myths"*. Marine Biology. National Geographic. Archived from *the original* on 26 April 2009. *Retrieved 2 May 2010.*

19. ^ Jump up to:[ab] ISAF Statistics on Attacking Species of Shark

20. ^ rice, doyle. *"2020 was an 'unusually deadly year' for shark attacks, with the most deaths since 2013"*. usa today.

21. ^ Jump up to:[ab]*"Carcharodon carcharias,Great white"*. FishBase.

22. ^ Jump up to:[ab]*"Family Lamnidae - Mackerel sharks or white shark"*. FishBase.

23. ^*"Common names of Carcharodon carcharias"*. FishBase.

24. ^ Martins, C.; Knickle, C. (2018). *"Carcharodon carcharias"*. Florida Museum.

25. ^ Jump up to:[ab] Martin, R. A. *"White Shark orGreat white?"*. Elasmo Research.

26. ^ Jump up to:[ab] Linnaeus, C (1758). *Systema naturae per regna tria naturae, secundum classes, ordines, genera, species, cum characteribus, differentiis, synonymis, locis. Tomus I. Editio decima, reformata* (in Latin). Vol. 1. Holmiae (Laurentii Salvii). p. 235. *doi:10.5962/ bhl.title.542.*

27. ^ Jordan, D. S. (1925). "The Generic Name of theGreat white, Squalus carcharias L.". Copeia. **140** (1925): 17–20. *doi:10.2307/1435586*. *JSTOR1435586*.

28. ^ Jump up to:[ab] Costantino, G. (18 August 2014). *"Sharks Were Once Called Sea Dogs, And Other Little-Known Facts"*. Smithsonain.com. Smithsonian Institution. *Retrieved 18 May 2019.*

29. ^ Human, B. A.; Owen, E. P.; Compagno, L. J. V.; Harley,

E. H. (2006). "Testing morphologically based phylogenetic theories within the cartilaginous fishes with molecular data, with special reference to the catshark family (Chondrichthyes; Scyliorhinidae) and the interrelationships within them". Molecular Phylogenetics and Evolution. **39** (2006): 384–391. *doi*:*10.1016/j.ympev.2005.09.009*. *PMID16293425*.

30. ^ Jump up to:*ab* Martin, A. P. (1996). "Systematics of the Lamnidae and the Origination Time of Carcharodon carcharias Inferred from the Comparative Analysis of Mitochondrial DNA Sequences". In Klimley, A. P.; Ainley, D. G. (eds.).Great whites: The Biology of Carcharodon carcharias. Academic Press. pp. 49–53. *doi*:*10.1016/B978-0-12-415031-7.X5000-9*. *ISBN978-0-12-415031-7*.

31. ^ Jump up to:*abcdefg* Kent, B. W. (2018). "The Cartilaginous Fishes (Chimaeras, Sharks, and Rays) of Calvert Cliffs, Maryland, USA". In Godfrey, S. J. (ed.). *The Geology and Vertebrate Paleontology of Calvert Cliffs, Maryland*. Smithsonian Contributions to Paleobiology. The Smithsonian Institution. pp. 45–157. *doi*:*10.5479/ si.1943-6688.100*. *ISSN1943-6688*. *S2CID134274604*.

32. ^ Jump up to:*ab* Applegate, S. P.; Espinosa-Arrubarrena, L. (1996). "The Fossil History of Carcharodon and Its Possible Ancestor, Cretolamna: A Study in Tooth Identification". In Klimley, A. P.; Ainley, D. G. (eds.).Great whites: The Biology of Carcharodon carcharias. Academic Press. pp. 19–36, 49–53. *doi*:*10.1016/B978-0-12-415031-7.X5000-9*. *ISBN978-0-12-415031-7*.

33. ^ Shimada, K. (2019). "The size of the megatooth shark, Otodus megalodon (Lamniformes: Otodontidae), revisited". Historical Biology. **33** (7): 1–8. *doi*:*10.1080/*

08912963.2019.1666840. ISSN0891-2963. S2CID208570844.

34. ⌃ Cooper, J. A.; Pimiento, C.; Ferrón, H. G.; Benton, M. J. (2020). _"Body dimensions of the extinct giant shark Otodus megalodon: a 2D reconstruction"_. Scientific Reports. **10** (14596): 14596. _Bibcode_:2020NatSR..1014596C. _doi_:10.1038/s41598-020-71387-y. PMC7471939. PMID32883981.

35. ^ Jump up to:[abcde] Ehret, D. A.; MacFadden, B. J.; Jones, D.; DeVries, T. J.; Foster, D. A.; Salas-Gismondi, R. (2012). "Origin of the White Shark Carcharodon (Lamniformes: Lamnidae), Based on Recalibration of the Late Neogene, Pisco Formation of Peru". Palaeontology. **55** (6): 1139–1153. _doi_:10.1111/j.1475-4983.2012.01201.x.

36. ⌃ Nyberg, K. G.; Ciampaglio, C. N.; Wray, G. A. (2006). "Tracing the Ancestry of theGreat white, Carcharodon carcharias, Using Morphometric Analyses of Fossil Teeth". Journal of Vertebrate Paleontology. **26** (4): 806–814. _doi_:10.1671/0272-4634(2006)26[806:ttaotg]2.0.co;2.

37. ⌃ Boessenecker, R. W.; Ehret, D. J.; Long, D. J.; Churchill, M.; Martin, E.; Boessenecker, S. J. (2019). _"The Early Pliocene extinction of the mega-toothed shark Otodus megalodon: a view from the eastern North Pacific"_. PeerJ. **7**: e6088. _doi_:10.7717/peerj.6088. PMC6377595. PMID30783558.

38. ⌃ Long, D. J.; Boessenecker, R. W.; Ehret, D. J. (2014), _Timing of evolution in the Carcharodon lineage: Rapid morphological change creates a major shift in a predator's trophic niche_

39. ⌃ Ebersole, J.A.; Ebersole, S.M.; Cicimurri, D.J. (2017). _"The occurrence of early Pleistocene marine fish remains_

from the Gulf Coast of Mobile County, Alabama". Palaeodiversity. **10** (1): 97–115. *doi*:*10.18476/ pale.v10.a6*. *S2CID134476316*.

40. ^ Jump up to:[ab] Yun, C. (2021). *"A tooth of the extinct lamnid shark, Cosmopolitodus planus comb. noc. (Chondrichthyes, Elasmobranchii) from the Miocene of Pohang City, South Korea"(PDF)*. Acta Palaeontologica Romaniae. **18** (1): 9–16. *doi*:*10.35463/j.apr.2022.01.02*. *S2CID242113412*.

41. ^ Jump up to:[ab] Martin, R. A. *"Fossil History of the White Shark"*. Elasmo Research.

42. ^ Trif, N.; Ciobanu, R.; Codrea, V. (2016). *"The first record of the giant shark Otodus megalodon (Agassiz, 1835) from Romania"*. Brukenthal, Acta Musei. **11** (3): 507–526.

43. ^*"Areal Distribution of the White Shark"*. National Capital Freenet. Archived from *the original* on 10 October 2018. *Retrieved 16 October 2010.*

44. ^ Kabasakal, H. (2014). *"The status of theGreat white (Carcharodon carcharias) in Turkey's waters"(PDF)*. Marine Biodiversity Records. **7**. *doi*:*10.1017/ S1755267214000980*.

45. ^*"Seabird predation by white shark, Carcharodon carcharias, and Cape fur seal, Arctocephalus pusillus pusillus, at Dyer Island"*. South African Journal of Wildlife Research. *Retrieved 22 May 2017.*

46. ^*"South Africa – Australia – South Africa"*. White Shark Trust.

47. ^ Thomas, Pete (29 September 2006). *"The Great White Way"*. Los Angeles Times. *Retrieved 1 October 2006.*

48. ^ Curtis, Tobey; McCandless, Camilla; Carlson, John; Skomal, Gregory; Kohler, Nancy; Natanson, Lisa; Burgess, George; Hoey, John; Pratt, Harold (June 2014).

"Seasonal Distribution and Historic Trends in Abundance of White Sharks, Carcharodon carcharias, in the Western North Atlantic Sea"(PDF). PLOS ONE. **9** (6): 12. *Bibcode:2014PLoSO...999240C*. *doi:10.1371/journal.pone.0099240*. *PMC4053410*. *PMID24918579*.

49. ^ Fieldstadt, Elisha (2 August 2019). *"More than 150Great whites sightings logged off Cape Cod, Massachusetts, since June"*. NBC News. *Retrieved 5 August 2019.*

50. ^ Jump up to:*ab* Wasser, Miriam (2 August 2019). *"Seals On Cape Cod Are More Than Just Shark Bait"*. WBUR. *Retrieved 5 August 2019.*

51. ^ Annear, Steve (18 June 2019). *"TrackingGreat whites off Cape Cod could help protect beachgoers"*. The Boston Globe. *Retrieved 5 August 2019.*

52. ^*"Atlantic White Sharks Research — AWSC"*. Atlantic White Shark Conservancy. *Retrieved 29 August 2021.*

53. ^ Finucane, Martin (23 June 2018). *"Great white sharks like to hang out in sea eddies, new study says"*. *The Boston Globe*. Archived from *the original* on 24 June 2018. *Retrieved 23 June 2018.*

54. ^*"Confirmed: 'Albino'Great white washes up in Australia | Sharks | Earth Touch News"*. Earth Touch News Network.

55. ^ Jump up to:*ab"Great White Sharks, Carcharodon carcharias"*. Marine Bio. *Retrieved 20 August 2012.*

56. ^*"Great White Sharks Have Blue Eyes"*. The Atlantic White Shark Conservancy. Archived from *the original* on 2 July 2018. *Retrieved 2 July 2018.*

57. ^ Echenique, E. J. *"A Shark to Remember: The Story of a Great White Caught in 1945"*. Archived from *the original* on 27 April 2012. *Retrieved 22 January 2013.* Home Page of Henry F. Mollet, Research Affiliate, Moss Landing Marine Laboratories.

58. ^ Jump up to:*abc* Tricas, T. C.; McCosker, J. E. (12 July 1984). *"Predatory behaviour of the white shark (Carcharodon carcharias), with notes on its biology"*. Proceedings of the California Academy of Sciences. *California Academy of Sciences*. **43** (14): 221–238. *Retrieved 22 January 2013*.

59. ^ Jump up to:*abc* Wood, Gerald (1983). *The Guinness Book of Animal Facts and Feats*. *ISBN978-0-85112-235-9*.

60. ^ Jump up to:*abc*Ellis, Richard and John E. McCosker. 1995.Great white. Stanford University Press, ISBN0-8047-2529-2

61. ^*"ADW: Carcharodon carcharias: Information"*. Animal Diversity Web. *Retrieved 16 May 2016*.

62. ^ Jump up to:*abcd* Cappo, Michael (1988). *"Size and age of the white pointer shark, Carcharodon carcharias (Linnaeus)"*. SAFISH. **13** (1): 11–13. Archived from *the original* on 6 January 2007.

63. ^ Taylor, Leighton R. (1 January 1993). *Sharks of Hawaii: Their Biology and Cultural Significance*. University of Hawaii Press. p. 65. *ISBN978-0-8248-1562-2*.

64. ^ Jump up to:*abc* Wroe, S.; Huber, D. R.; Lowry, M.; McHenry, C.; Moreno, K.; Clausen, P.; Ferrara, T. L.; Cunningham, E.; Dean, M. N.; Summers, A. P. (2008). *"Three-dimensional computer analysis of white shark jaw mechanics: how hard can a great white bite?"(PDF)*. Journal of Zoology. **276** (4): 336–342. *doi:10.1111/j.1469-7998.2008.00494.x*.

65. ^*"Great White Shark"*. National Geographic. *Retrieved 24 July 2010*.

66. ^ Jump up to:*ab* Mollet, H. F. (2008), *White Shark Summary Carcharodon carcharias (Linnaeus, 1758)]*,

Home Page of Henry F. Mollet, Research Affiliate, Moss Landing Marine Laboratories, archived from *the original* on 31 May 2012

67. ^ Jump up to:*abcdefghijklmn* Klimley, Peter; Ainley, David (1996).Great whites: The Biology of Carcharodon carcharias. Academic Press. pp. 91–108. *ISBN978-0-12-415031-7*.

68. ^ Jury, Ken (1987). *"Huge 'White Pointer' Encounter"*. SAFISH. **2** (3): 12–13. Archived from *the original* on 17 April 2012.

69. ^*"Ep. 10. A Bathing Accident - Transcript"*. Shark Files.

70. ^*"Learn More About Deep Blue, One of the BiggestGreat whites Ever Filmed"*. Discovery.

71. ^ Staff, H. N. N. *"BiggestGreat white on record seen in Hawaii waters"*. Hawaii News.

72. ^ Ciaccia, Chris (16 January 2019). *"Great white shark, called 'Deep Blue,' spotted near Hawaii"*. Fox News.

73. ^*"Deep Blue, perhaps the largest knownGreat white, spotted off Hawaii"*. 16 January 2019.

74. ^ Armitage, Stefan (12 July 2019). *"Fisherman encounters '30-foot'Great white"*. Vt.co. *Retrieved 28 November 2019.*

75. ^ Ferreira, Craig (2011).Great whites On Their Best Behavior.

76. ^ Froese, Rainer; Pauly, Daniel (eds.) (2011). *"Galeocerdo cuvier"* in FishBase. July 2011 version.

77. ^*"Summary of Large Tiger Sharks Galeocerdo cuvier (Peron & LeSueur, 1822)"*. Archived from *the original* on 10 April 2012. *Retrieved 3 May 2010.*

78. ^ Eagle, Dane. *"Greenland Shark"*. Florida Museum of Natural History. Archived from *the original* on 5 April 2013. *Retrieved 1 September 2012.*

79. ^ Martin, R. Aidan. *"Pacific Sleeper Shark"*. ReefQuest

Centre for Shark Research. Biology of Sharks and Rays. Archived from *the original* on 20 April 2013. *Retrieved 1 September 2012.*

80. ^ Jump up to:[abc]*"LargestGreat white Specimens by Paleonerd01 on DeviantArt"*. www.deviantart.com.

81. ^ Jump up to:[abc] Christiansen, Heather M.; Lin, Victor; Tanaka, Sho; Velikanov, Anatoly; Mollet, Henry F.; Wintner, Sabine P.; Fordham, Sonja V.; Fisk, Aaron T.; Hussey, Nigel E. (16 April 2014). *"The Last Frontier: Catch Records of White Sharks (Carcharodon carcharias) in the Northwest Pacific Sea"*. PLOS ONE. **9** (4): e94407. *Bibcode*:*2014PLoSO...994407C*. *doi*:*10.1371/ journal.pone.0094407*. *PMC3989224*. *PMID24740299*.

82. ^*"Barnacle Lil' Still Terror Of West Coast Deep"*. *The Chronicle*. *Streaky Bay*. 17 April 1952 – via *Trove*.

83. ^*"White Shark Summary Carcharodon carcharias (Linnaeus, 1758)"*. www.elasmollet.org.

84. ^ Alessandro De Maddalena & Walter Heim (2015). MediterraneanGreat whites: A Comprehensive Study Including All Recorded Sightings. McFarland. *ISBN978-0786488155*.

85. ^*"On theGreat white, Carcharodon carcharias (Linnaeus, 1758), preserved in the Museum of Zoology in Lausanne"*. ResearchGate.

86. ^ Jump up to:[ab] Nevres, M. Özgür (6 October 2018). *"LargestGreat whites ever recorded"*. Our Planet.

87. ^*"An analysis of photographic evidences of the largestGreat whites (Carcharodon carcharias), Linnaeus 1758, captures in the Mediterranean sea with considerations about the maximum size of the species"*. ResearchGate.

88. ^*"Squalus Carcharias: Le TRÈS Grand Requin Blanc !"*. 29 July 2013.

89. ^ Jump up to:_ab_Gerald L Wood (1982). Guinness Book of World Records Animal Facts and Feats. Sterling Pub Co., Inc. _ISBN978-0851122359_.

90. ^ Molotsky, Irvin M (24 June 1978). _"Harpooned Shark Pulls Fishing Boat 14 Hours Off L.I."New York Times_.

91. ^_"Deep Blue"_. 19 March 2017.

92. ^_"One of biggestGreat whites seen feasting on sperm whale in rare video"_. Animals. 19 July 2019.

93. ^_"The Big One, a Great White story from Grand Manan"_. new-brunswick.net.

94. ^ Vladykov, Vadim D. (Vadim Dmitrij); McKenzie, Ross A. (17 May 1935). _"The marine fishes of Nova Scotia"_. Proceedings of the Nova Scotian Institute of Science – via dalspace.library.dal.ca.

95. ^_"Photograph, Shark at Cowes, 1987"_. Victorian Collections, Phillip Island and District Historical Society.

96. ^_"Photo of dead shark"_.

97. ^ King B, Hu Y, Long JA (May 2018). _"Electroreception in early vertebrates: survey, evidence and new information"_. Palaeontology. **61** (3): 325–58. _doi_:_10.1111/pala.12346_.

98. ^_"The physiology of the ampullae of Lorenzini in sharks"_. Biology Dept., Davidson College. Biology @ Davidson. Archived from _the original_ on 24 November 2010. _Retrieved 20 August 2012_.

99. ^ Martin, R. Aidan. _"Body Temperature of the Great white and Other Lamnoid Sharks"_. ReefQuest Centre for Shark research. _Retrieved 16 October 2010_.

100. ^White Shark Biological ProfileArchived 20 January 2013 at WebCite from the Florida Museum of Natural History

101. ^ Barber, Elizabeth (18 July 2013). _"Great white shark packs its lunch in its liver before a big trip"_. The Christian

Science Monitor. Archived from *the original* on 2 August 2013.

102. ^ Jordan, Rob (17 July 2013). *"Great White Sharks' Fuel for Oceanic Voyages: Liver Oil"*. sciencedaily.com. Stanford University.

103. ^ Solly, Meilan (3 April 2019). *"Great White Sharks Thrive Despite Heavy Metals Coursing Through Their Veins"*. Smithsonian Magazine. *Archived* from the original on 30 June 2019. *Retrieved 30 June 2019.*

104. ^ Medina, Samantha (27 July 2007). *"Measuring the great white's bite"*. Cosmos Magazine. Archived from *the original* on 5 May 2012. *Retrieved 1 September 2012.*

105. ^ Martin, R. Aidan; Martin, Anne (October 2006). *"Sociable Killers: New studies of the white shark (aka great white) show that its social life and hunting strategies are surprisingly complex"*. *Natural History*. *Retrieved 24 November 2019.*

106. ^ Papastamatiou Yannis P., Mourier Johann, TinHan Thomas, Luongo Sarah, Hosoki Seiko, Santana-Morales Omar and Hoyos-Padilla Mauricio. 2022 Social dynamics and individual hunting tactics of white sharks revealed by biologging. Biol. Lett.18: 2021059920210599 http://doi.org/10.1098/rsbl.2021.0599

107. ^ Jump up to:[abcde] Martin, R. Aidan; Martin, Anne. *"Sociable Killers"*. Natural History Magazine. Archived from *the original* on 15 May 2013. *Retrieved 30 September 2006.*

108. Johnson, R. L.; Venter, A.; Bester, M.N.; Oosthuizen, W.H. (2006). *"Seabird predation by white shark Carcharodon carcharias and Cape fur seal Arctocephalus pusillus pusillus at Dyer Island"(PDF)*. South African Journal of Wildlife Research. South Africa. **36** (1):

23–32. Archived from *the original*(PDF) on 3 April 2012.

109. ^TeenageGreat whites are awkward biters. *Science Daily* (2 December 2010)

110. ^White shark diets show surprising variability, vary with age and among individuals. *Science Daily* (29 September 2012)

111. ^ Estrada, J. A.; Rice, Aaron N.; Natanson, Lisa J.; Skomal, Gregory B. (2006). "Use of isotopic analysis of vertebrae in reconstructing ontogenetic feeding ecology in white sharks". Ecology. **87** (4): 829–834. *doi*:*10.1890/ 0012-9658(2006)87[829:UOIAOV]2.0.CO;2*. *PMID16676526*.

112. ^ Jump up to:*ab* Fergusson, I. K.; Compagno, L. J.; Marks, M. A. (2000). "Predation by white sharks Carcharodon carcharias (Chondrichthyes: Lamnidae) upon chelonians, with new records from the Mediterranean Sea and a first record of the sea sunfish Mola mola (Osteichthyes: Molidae) as stomach contents". Environmental Biology of Fishes. **58** (4): 447–453. *doi*:*10.1023/a:1007639324360*. *S2CID31232421*.

113. ^ Hussey, N. E., McCann, H. M., Cliff, G., Dudley, S. F., Wintner, S. P., & Fisk, A. T. (2012). Size-based analysis of diet and trophic position of the white shark (*Carcharodon carcharias*) in South African waters. *Global Perspectives on the Biology and Life History of the White Shark.* (Ed. ML Domeier.) pp. 27–49.

114. ^*"Catch as Catch Can"*. ReefQuest Centre for Shark Research. *Retrieved 16 October 2010.*

115. ^*"How Fast Can a Shark Swim?"*. ReefQuest Centre for Shark Research.

116. ^*"White Shark Predatory Behavior at Seal Island"*. ReefQuest Centre for Shark Research.

117. ^ Le Boeuf, B. J.; Crocker, D. E.; Costa, D. P.; Blackwell, S. B.; Webb, P. M.; Houser, D. S. (2000). "Foraging ecology of northern elephant seals". Ecological Monographs. **70** (3): 353–382. *doi*:*10.2307/2657207*. *JSTOR2657207*.

118. ^ Haley, M. P.; Deutsch, C. J.; Le Boeuf, B. J. (1994). "Size, dominance and copulatory success in male northern elephant seals, Mirounga angustirostris". Animal Behaviour. **48** (6): 1249–1260. *doi*:*10.1006/ anbe.1994.1361*. *S2CID54388167*.

119. ^ Weng, K. C.; Boustany, A. M.; Pyle, P.; Anderson, S. D.; Brown, A.; Block, B. A. (2007). "Migration and habitat of white sharks (Carcharodon carcharias) in the eastern Pacific Sea". Marine Biology. **152** (4): 877–894. *doi*:*10.1007/s00227-007-0739-4*. *S2CID39985022*.

120. ^ Martin, Rick. *"Predatory Behavior of Pacific Coast White Sharks"*. Shark Research Committee. Archived from *the original* on 30 July 2012.

121. ^ Chewning, Dana; Hall, Matt. *"Carcharodon carcharias (Great white shark)"*. Animal Diversity Web. *Retrieved 15 August 2019.*

122. ^ Jump up to:[abcdef] Heithaus, Michael (2001). *"Predator–prey and competitive interactions between sharks (order Selachii) and dolphins (suborder Odontoceti): a review"(PDF)*. Journal of Zoology. **253**: 53–68. *CiteSeerX10.1.1.404.130*. *doi*:*10.1017/ S0952836901000061*. Archived from *the original(PDF)* on 15 January 2016. *Retrieved 26 February 2010.*

123. ^ Long, Douglas (1991). *"Apparent Predation by a White Shark Carcharodon carcharias on a Pygmy Sperm Whale Kogia breviceps"(PDF)*. Fishery Bulletin. **89**: 538–540.

124. ^ Kays, R. W., & Wilson, D. E. (2009). *Mammals of North America*. Princeton University Press.

125. ^ Baird, R. W.; Webster, D. L.; Schorr, G. S.; McSweeney, D. J.; Barlow, J. (2008). _"Diet variation in beaked whale diving behavior"_(PDF). Marine Mammal Science. **24** (3): 630–642. _doi:10.1111/j.1748-7692.2008.00211.x_. _hdl:10945/697_.

126. ^ Jump up to:_ab_ Krkosek, Martin; Fallows, Chris; Gallagher, Austin J.; Hammerschlag, Neil (2013). _"White Sharks (Carcharodon carcharias) Scavenging on Whales and Its Potential Role in Further Shaping the Ecology of an Apex Predator"_. PLOS ONE. **8** (4): e60797. _Bibcode:2013PLoSO...860797F_. _doi:10.1371/journal.pone.0060797_. _PMC3621969_. _PMID23585850_.

127. ^ Dudley, Sheldon F. J.; Anderson-Reade, Michael D.; Thompson, Greg S.; McMullen, Paul B. (2000). _"Concurrent scavenging off a whale carcass byGreat whites, Carcharodon carcharias, and tiger sharks, Galeocerdo cuvier"_(PDF). Marine Biology. Fishery Bulletin. Archived from _the original_(PDF) on 27 May 2010. _Retrieved 4 May 2010._

128. ^ Dines, Sasha; Gennari, Enrico (29 January 2020). _"First observations of white sharks (Carcharodon carcharias) attacking a live humpback whale (Megaptera novaeangliae)"_. Marine and Freshwater Research. **71** (9): 1205. _doi:10.1071/MF19291_. _S2CID212969014_ – via www.publish.csiro.au.

129. ^ Márquez, Melissa Cristina. _"First Observations Of White Sharks Attacking A Live Humpback Whale"_. Forbes.

130. ^_"Drone footage shows aGreat white drowning a 33ft humpback whale"_. The Independent. 15 July 2020.

131. ^ Fish, Tom (15 July 2020). _"Shark attack: Watch 'strategic' Great White hunt down and kill 10 Metre humpback whale"_. Express.co.uk.

132. ⌃ Owens, Brian (12 February 2016). *"White shark's diet may include biggest fish of all: whale shark"*. New Scientist.
133. ⌃ Moore, G. I.; Newbrey, M. G. (2015). "Whale shark on a white shark's menu". Marine Biodiversity. **46** (4): 745. *doi*:*10.1007/s12526-015-0430-9*. *S2CID36426982*.
134. ⌃*"Natural History of the White Shark"*. PRBO Conservation Science. 2 May 2010. Archived from *the original* on 3 July 2013.
135. ⌃*"LegendaryGreat white Was Just A Teenager When Killed, New Research Reveals"*. The Inquisitr News. 9 March 2015.
136. ⌃ Roy, Eleanor Ainge (4 September 2020). *"'Rolling and rolling and rolling': the first detailed account ofGreat white sex"* – via www.theguardian.com.
137. ⌃*"Carcharodon carcharias,Great whites"*. marinebio.org.
138. ⌃*"Brief Overview of theGreat white (Carcharodon carcharias)"*. Elasmo Research. *Retrieved 20 August 2012.*
139. ⌃ Animals, Kimberly Hickok 2019-03-22T19:03:40Z (22 March 2019). *"EnormousGreat white Pregnant with Record 14 Pups Was Caught and Sold in Taiwan"*. livescience.com.
140. ⌃ Martin, R. Aidan. *"White Shark Breaching"*. ReefQuest Centre for Shark Research. *Retrieved 18 April 2012.*
141. ⌃ Martin, R. A.; Hammerschlag, N.; Collier, R. S.; Fallows, C. (2005). "Predatory behaviour of white sharks (Carcharodon carcharias) at Seal Island, South Africa". Journal of the Marine Biological Association of the UK. **85** (5): 1121. *CiteSeerX10.1.1.523.6178*. *doi*:*10.1017/S002531540501218X*. *S2CID17889919*.
142. ⌃ Rice, Xan (19 July 2011). *"Great white shark jumps from sea into research boat"*. *The Guardian*. London.

Retrieved 20 July 2011. Marine researchers in South Africa had a narrow escape after a 3 m (10 ft) longGreat white breached the surface of the sea and leapt into their boat, becoming trapped on deck for more than an hour. [...] Enrico Gennari, an expert onGreat whites, [...] said it was almost certainly an accident rather than an attack on the boat.

143. ^ Jump up to:*ab* Pyle, Peter; Schramm, Mary Jane; Keiper, Carol; Anderson, Scot D. (26 August 2006). *"Predation on a white shark (Carcharodon carcharias) by a killer whale (Orcinus orca) and a possible case of competitive displacement"(PDF)*. Marine Mammal Science. **15** (2): 563–568.   *doi:10.1111/j.1748-7692.1999.tb00822.x*. Archived from *the original(PDF)* on 22 March 2012. *Retrieved 8 May 2010.*

144. ^ Jump up to:*ab"Nature Shock Series Premiere: The Whale That Ate the Great White"*. Tvthrong.co.uk. 4 October 1997. Archived from *the original* on 6 April 2012. *Retrieved 16 October 2010.*

145. ^*"Killer Whale Documentary Part 4"*. youtube.com. Archived from *the original* on 25 July 2013.

146. ^ Jump up to:*abc* Turner, Pamela S. (October–November 2004). *"Showdown at Sea: What happens whenGreat whites go fin-to-fin with killer whales?"*. National Wildlife. *National Wildlife Federation*. **42** (6). Archived from *the original* on 16 January 2011. *Retrieved 21 November 2009.*

147. ^*"Great white shark 'slammed' and killed by a pod of killer whales in South Australia"*. Australian Broadcasting Corporation. 3 February 2015. *Retrieved 10 July 2015.*

148. ^ Haden, Alexis (6 June 2017). *"Killer whales have been killingGreat whites in Cape waters"*. The South African. *Retrieved 27 June 2017.*

149. ^ Jorgensen, S. J.; et al. (2019). *"Killer whales redistribute white shark foraging pressure on seals"*. Scientific Reports. **9** (1): 6153. *Bibcode*:*2019NatSR...9.6153J*. *doi*:*10.1038/s41598-019-39356-2*. *PMC6467992*. *PMID30992478*.

150. ^ Starr, Michell (11 November 2019). *"Incredible Footage Reveals Orcas Chasing Off The Sea's Most Terrifying Predator"*. Science Alert. *Retrieved 24 November 2019*.

151. ^ Benchley, Peter (April 2000). "Great white sharks". *National Geographic*: 12. *ISSN0027-9358*. considering the knowledge accumulated about sharks in the last 25 years, I couldn't possibly write Jaws today ... not in good conscience anyway ... back then, it was OK to demonize an animal.

152. ^*"Great White Shark Attacks: Defanging the Myths"*. nationalgeographic.com. 23 January 2004.

153. ^ Martin, R. Aidan (2003). *"White Shark Attacks: Mistaken Identity"*. Biology of Sharks and Rays. ReefQuest Centre for Shark Research. *Retrieved 30 August 2016*.

154. ^ Ryan, Laura A.; Slip, David J.; Chapuis, Lucille; Collin, Shaun P.; Gennari, Enrico; Hemmi, Jan M.; How, Martin J.; Huveneers, Charlie; Peddemors, Victor M.; Tosetto, Louise; Hart, Nathan S. (2021). *"A shark's eye view: testing the 'mistaken identity theory' behind shark bites on humans"*. Journal of the Royal Society Interface. **18** (183): 20210533. *doi*:*10.1098/rsif.2021.0533*. *PMC* 8548079. *PMID34699727* – via royalsocietypublishing.org (Atypon).

155. ^*"ISAF Statistics for Worldwide Unprovoked White Shark Attacks Since 1990"*. 10 February 2011. *Retrieved 19 August 2011*.

156. ^ Tricas, T.C.; McCosker, John (1984). *"Predatory behavior of the white shark, Carcharodon carcharias, and*

notes on its biology". Proceedings of the California Academy of Sciences. Series 4. **43** (14): 221–238.

157. ^ Jump up to:[abcdefg] *"Shark culling"*. marineconservation.org.au. Archived from *the original* on 2 October 2018. *Retrieved 30 August 2019.*

158. ^ Phillips, Jack (4 September 2018), *Video: Endangered Hammerhead Sharks Dead on Drum Line in Great Barrier Reef*, Ntd.tv, archived from *the original* on 19 September 2018, *retrieved 30 August 2019*

159. ^ Thom Mitchell (20 November 2015), *Action for Dolphins. Queensland's Shark Control Program Has Snagged 84,000 Animals*, *retrieved 30 August 2019*

160. ^ Jump up to:[ab] Roff, George; Brown, Christopher J.; Priest, Mark A.; Mumby, Peter J. (2018). *"Decline of coastal apex shark populations over the past half century"*. Communications Biology. **1** (1): 1–11. *doi*:*10.1038/s42003-018-0233-1*. *PMC6292889*. *PMID30564744*.

161. ^ Jump up to:[ab] Wang, Kelly (4 September 2018). *"Heartbreaking Photos Show the Brutal Lengths Australia Is Going to In Order to 'Keep Sharks Away From Tourists'"*. One Green Planet. *Retrieved 29 August 2019.*

162. ^ Mackenzie, Bruce (4 August 2018), *Sydney Shark Nets Set to Stay Despite Drumline Success*, Swellnet.com., *retrieved 30 August 2019*

163. ^*Shark Nets*, Sharkangels.org, archived from *the original* on 19 September 2018, *retrieved 30 August 2019*

164. ^*"New measures to combat WA shark risks"*. Department of Fisheries, Western Australia. 10 December 2013. Archived from *the original* on 1 February 2014. *Retrieved 2 February 2014.*

165. ^ Arup, Tom (21 January 2014), *"Greg Hunt grants WA exemption for shark cull plan"*, The Sydney Morning Herald, Fairfax Media, archived from *the original* on 22

January 2014

166. ^*"Can governments protect people from killer sharks?"*. Australian Broadcasting Corporation. 22 December 2013. *Retrieved 2 February 2014.*

167. ^Australia shark policy to stay, despite threats*TVNZ*, 20 January 2014.

168. ^*"More than 100 shark scientists, including me, oppose the cull in Western Australia"*. 23 December 2013. *Retrieved 31 August 2016.*

169. ^*"Unprovoked White Shark Attacks on Kayakers"*. Shark Research Committee. *Retrieved 14 September 2008.*

170. ^ Tricas, Timothy C.; McCosker, John E. (1984). *"Predatory Behaviour of the White Shark (Carcharodon carcharias), with Notes on its Biology"(PDF)*. Proceedings of the California Academy of Sciences. **43** (14): 221–238.

171. ^*"Great white shark sets record at California aquarium"*. *USA Today*. 2 October 2004. *Retrieved 27 September 2006.*

172. ^ Hopkins, Christopher Dean (8 January 2016). *"Great White Shark Dies After Just 3 Days In Captivity At Japan Aquarium"*. NPR. *Archived* from the original on 3 April 2017. *Retrieved 21 December 2017.*

173. ^ Gathright, Alan (16 September 2004). *"Great white shark puts jaws on display in aquarium tank"*. San Francisco Chronicle. *Retrieved 27 September 2006.*

174. ^*"White Shark Research Project"*. *Monterey Bay Aquarium*. Archived from *the original* on 19 January 2013. *Retrieved 27 September 2006.*

175. ^ Squatriglia, Chuck (1 September 2003). *"Great white shark introduced at Monterey Bay Aquarium"*. San Francisco Chronicle. *Retrieved 27 September 2006.*

176. ^*"Learn All About Our New White Shark"*. Monterey Bay

Aquarium. Archived from *the original* on 20 November 2009. *Retrieved 28 August 2009.*

177. ^*"NewGreat white goes on display at Monterey Bay Aquarium"*. 1 September 2011.

178. ^*"Great White Shark Dies Shortly After Release From Monterey Aquarium"*. 3 November 2011.

179. ^*"Great white shark dies after release from Monterey Bay Aquarium"*. Los Angeles Times. 3 November 2011.

180. ^*"White Shark"*. Monterey Bay Aquarium. *Retrieved 21 August 2021.*

181. ^ Hongo, Jun (8 January 2016). *"Great White Shark Dies at Aquarium in Japan"*. Wall Street Journal. *Retrieved 9 January 2016.*

182. ^*"Great white shark dies after three days in Japanese aquarium"*. Telegraph.co.uk. Archived from *the original* on 7 January 2016. *Retrieved 9 January 2016.*

183. ^*"Electroreception"*. Elasmo-research. *Retrieved 27 September 2006.*

184. ^*"There's a Reason You'll Never See aGreat white in an Aquarium"*.

185. ^*"Shark cage diving"*. Department of Environment, Water and Natural Resources. Archived from *the original* on 9 April 2013. *Retrieved 11 March 2013.*

186. ^ Squires, Nick (18 January 1999). *"Swimming With Sharks"*. BBC. Archived from *the original* on 17 August 2003. *Retrieved 21 January 2010.*

187. ^ Simon, Bob (11 December 2005). *"Swimming With Sharks"*. *60 Minutes*. *Retrieved 22 January 2010.*

188. ^ Jump up to:[ab]*"Blue Water Hunting Successfully"*. Blue Water Hunter. Archived from *the original* on 18 August 2012. *Retrieved 20 August 2012.*

189. ^*"AGreat white's favorite tune? 'Back in Black'"* Archived 16 April 2016 at the Wayback Machine*Surfersvillage*

Global Surf News (3 June 2011). Retrieved 30 January 2014.

190. ^*"Shark Attacks Compared to Lightning"*. *Florida Museum of Natural History*. 18 July 2003. *Retrieved 7 November 2006.*

191. ^ Hamilton, Richard (15 April 2004). *"SA shark attacks blamed on tourism"*. BBC. Archived from *the original* on 23 March 2012. *Retrieved 24 October 2006.*

192. ^*"Regulation of Trade in Specimens of Species Included in Appendix II"*. CITES (1973). Archived from *the original* on 17 July 2011. *Retrieved 8 April 2012.*

193. ^*"Memorandum of Understanding on the Conservation of Migratory Sharks"(PDF)*. Convention on Migratory Species. 12 February 2010. Archived from *the original(PDF)* on 20 April 2013. *Retrieved 31 August 2012.*

194. ^ Sample, Ian (19 February 2010). *"Great white shark is more endangered than tiger, claims scientist"*. The Guardian. *Retrieved 14 August 2013.*

195. ^ Jenkins, P. Nash (24 June 2014). *"Beachgoers Beware: TheGreat white Population Is Growing Again"*. Time. *Retrieved 29 October 2014.*

196. ^ Gannon, Megan (20 June 2014). *"Great White Sharks Are Making a Comeback off US Coasts"*. livescience.com. *Retrieved 29 October 2014.*

197. ^ De Maddalena, Alessandro; Heim, Walter (2012). MediterraneanGreat whites: A Comprehensive Study Including All Recorded Sightings. McFarland. *ISBN978-0-7864-5889-9*.

198. ^ Jump up to:*abc* Government of Australia. *"Species Profile and Threats Database — Carcharodon carcharias—Great White Shark"*. *Retrieved 21 August 2013.*

199. ^ Jump up to:*ab* Environment Australia (2002). *White Shark (Carcharodon carcharias) Recovery Plan* (Report).

200. ^ Blower, Dean C.; Pandolfi, John M.; Bruce, Barry D.; Gomez-Cabrera, Maria del C.; Ovenden, Jennifer R. (2012). *"Population genetics of Australian white sharks reveals fine-scale spatial structure, transoceanic dispersal events and low effective population sizes"*. Marine Ecology Progress Series. **455**: 229–244. *Bibcode*:*2012MEPS..455..229B*. *doi*:*10.3354/meps09659*.

201. ^*"About the Campaign: Sea Shepherd Working Together With The Community To Establish Sustainable Solutions To Shark Bite Incidents"*. seashepherd.org. Archived from *the original* on 8 April 2018. *Retrieved 29 August 2019*.

202. ^ Hillary, Rich; Bradford, Russ; Patterson, Toby (8 February 2018). *"World-first genetic analysis reveals Aussie white shark numbers"*. The Conversation. *Retrieved 21 August 2021*.

203. ^*"Great white sharks to be protected"*. *The New Zealand Herald*. 30 November 2006. *Retrieved 30 November 2006*.

204. ^ Duffy, Clinton A. J.; Francis, Malcolm; Dunn, M. R.; Finucci, Brit; Ford, Richard; Hitchmough, Rod; Rolfe, Jeremy (2018). *Conservation status of New Zealand chondrichthyans (chimaeras, sharks and rays), 2016*(PDF). Wellington, New Zealand: Department of Conservation. p. 9. *ISBN978-1-988514-62-8*. *OCLC1042901090*.

205. ^ Quan, Kristene (4 March 2013). *"Great White Sharks Are Now Protected under California Law"*. Time.

206. ^ Williams, Lauren (3 July 2014). "Shark numbers not tanking". Huntington Beach Wave. *The Orange County Register*. p. 12.

207. ^ Burgess, George H.; Bruce, Barry D.; Cailliet, Gregor

M.; Goldman, Kenneth J.; Grubbs, R. Dean; Lowe, Christopher J.; MacNeil, M. Aaron; Mollet, Henry F.; Weng, Kevin C.; O'Sullivan, John B. (16 June 2014). _"A Re-Evaluation of the Size of the White Shark (Carcharodon carcharias) Population off California, USA"_. _PLoS ONE_. 9 (6): e98078. _Bibcode:2014PLoSO...998078B_. _doi:10.1371/journal.pone.0098078_. _PMC4059630_. _PMID24932483_.

208. _^New Regulations Affecting Activity around White Sharks_. Commonwealth of Massachusetts, Division of Marine Fisheries (4 June 2015)

For Queries

For any queries, please send an email to- ath02mo@gmail.com